THE SOCIAL ORGANIZATION OF ELECTRIC POWER SUPPLY IN MODERN SOCIETIES

The MIT Press
Cambridge, Massachusetts, and London, England

THE SOCIAL ORGANIZATION OF ELECTRIC POWER SUPPLY IN MODERN SOCIETIES

Philip Sporn

This book was designed
by The MIT Press Design Department.
It was set in Linotype Baskerville
by The Colonial Press Inc.
printed on Warren's Old Style
by The Colonial Press Inc.
and bound in Interlaken Matte Yellow
by The Colonial Press Inc.
in the United States of America.

ISBN 0 262 190982 (hardcover)

Library of Congress catalog card number: 71-161661

To
Michael
Tom
Sarah
Paul
Robert
Abby

ACKNOWLEDGMENTS

I am glad to acknowledge and express my appreciation for the help I received from Mr. Abraham Gerber in the editing of some of the material and from my secretary, Miss Dorothy Miesse, both for help in the editing and for her work in getting the manuscript ready for the printer.

Philip Sporn

FOREWORD

The national search for a sound energy policy has been heightened by the environmental implications of a sixfold increase in energy requirements projected for the year 2000. Correspondingly, the debate between advocates of private and public power has taken on a new and added significance. How shall the nation seek to organize the generation, transmission, and distribution of electric power? Does the U. S. pattern of publicly regulated, investor-owned companies continue to represent an advantage over autonomous, publicly owned electric power companies? Where is the widespread concern over the public interest in environmental quality and the consumer interest in reliability and cost leading the balance between private and public power? What are the essential differences? How are the delicate balances among competing technologies changing the shape and form of this large and vital industry?

Philip Sporn, long an outspoken proponent of private investment in electric power, has examined these questions and other aspects of social and economic organization of the electric power industry in four lectures given at the Massachusetts Institute of Technology during the 1970–1971 academic year. This book is an effort by The MIT Press to bring these lectures to a wider audience.

Mr. Sporn has been in the vanguard of technological advance in the electric utility field for nearly fifty years. He has been aptly described by *Fortune* magazine as "the mass-producing, cost-cutting Henry Ford of electric power." No student of this country's progress in electric energy development can go far without meeting Philip Sporn—in the scholarly journals of electrical engineering or in the most advanced electric power systems installed anywhere in the world. As chief engineer of American Electric Power System and later as its president, he became the leading U. S. exponent of cheaper electric power through the relentless application of new technology. As a member of the National Academy of Sciences and the National Academy of Engineering, he has had more than a parochial view of the matter.

This book is Mr. Sporn's vision of the future of electric power in advanced societies. It is also a personal account which is best seen in the context of the man and the times. In these lectures, Mr. Sporn demonstrates the sensitivity to social and economic issues we can hope to encourage in our young engineers and in engineering curricula. He is on solid ground when he deplores the oversight of important costs in public power and at the center of controversy when he questions the ability of corporate directors to reflect the public interest adequately in privately managed companies. Students of biography and of industrial organization will find value in these lectures along with the public policy insights the author has sought to provide about the future of electric energy.

I am glad to have an opportunity to introduce this book by a long-standing friend and creative contributor to engineering education at M.I.T.

Howard W. Johnson

1

Lecture One

THE STATUS IN THE UNITED STATES; HOW IT IS DONE IN OTHER ECONOMICALLY ADVANCED COUNTRIES; IDEOLOGICAL CONSIDERATIONS

September 4, 1972, marks the ninetieth anniversary of the placing into service of the first commercial electric power station in the world, the famous Pearl Street Station in the city of New York.

With all the technological wonders we have witnessed since then—the automobile, the airplane, radio, the vacuum tube, television, transistors, digital computers, automation, nuclear energy, and space flights—one can still state that no invention or technical innovation since the establishment of that power station has had as much effect on man—his mastery of his environment, the raising of his productivity, his welfare, the enhancement of his enjoyment of living as a result of the free time thus made available to him, and his potentials for spiritual advancement—as has the development of electric power.

Lenin's famous dictum in 1920, "Communism is Soviet power plus electrification of the whole country," was perhaps not far from the mark.

In the United States at Pearl Street that afternoon of September 4, when the switch was thrown and commercial electric service was begun with a load of approximately 400 lamps, each taking about 83 watts and constituting an electric demand of 33 kilowatts, the modest display was the beginning of a fantastic development whose vitality has remained unabated to this day.

A year later the connected load had reached a total of about 860 kw. In 1970, the total noncoincident peak demand in the United States will be 275 million kw, and facilities that will be required to supply that demand will represent an investment of close to $100 billion.

The function of energy and of electric energy, as we commonly think of it, is to supply our society and its homes with all the countless necessities, comforts, and conveniences that have become almost indispensable elements of civilized living—lighting, heating, cooling, the daily chores of a home, cooking, cleaning, washing, entertainment by radio, television, home music, and home movies, and many other operations too numerous to mention. This function of energy holds equally true in business and in public life.

But the most important and specific function of energy is to operate the capital equipment of our mammoth industry. This equipment when applied to production processes increases productivity. This need for energy results from the marshaling and use of capital equipment in production. It is the availability of capital in the form of tools, machinery, and equipment that makes possible the production and high productivity of our American society. Yet none of this wide-flung industrial plant, extending over the length and breadth of this great nation, is capable of producing anything without the mobility that is furnished it by energy.

The position that electric energy has come to occupy in our civilization is a result of its unique characteristics of ease in transportation and distribution, ease of application to the widest range of processes requiring energy, and its potential for application to almost all energy-using processes with the utmost flexibility, complete cleanliness, and safety, and with the largest and most sensitive susceptibility to control.[1]

When electric energy started as a commercial enterprise, the concept of an energy supply as an indispensable utility in modern society had as yet not been born. Electric energy was just an industrial enterprise. By 1920, the United States had reached full stature as the world's foremost industrial and economically developed nation; electric energy had grown to the point where our total annual electric utility production was close to 40 billion kwh, of which almost 96% was produced by private enterprise and only slightly above 4% by government agencies—federal, state, municipal, and cooperatives. This ratio did not change much in the following decade.

Then in 1933 came the Tennessee Valley Authority (TVA), and the ownership picture started an accelerat-

[1] For more detailed discussion, see Philip Sporn, *Energy; Its Production, Conversion and Use in the Service of Man* (New York: Pergamon Press, 1963).

ing rate of change that continued unabated for two decades. Thus by 1950, with total electric generation having grown more than eightfold to 329 billion kwh, the governmental portion had grown to 19%, and the private or investor-owned part had declined to 81%. The governmental generation portion continued to increase, but at only a very moderate rate, from the level of just under 20% for the next 20 years and today has seemed to stabilize at around 23%.

Here in the United States, with less than 6% of the world's population, there has come into being the largest power industry in the world, furnishing the country with 35% of the world's total electric energy. On the whole, the record of the electric power industry in the United States is renowned the world over for quality and continuity. Nevertheless, any time there is a disturbance or malperformance, particularly such a deeply traumatic one as the New York–New England blackout of November 1965 (which had never happened before, should not have happened when it did, and can most certainly be avoided by well-known and thoroughly understood remedial practices in the future), the most panicky alarms of pending doom are heard. Almost immediately an inrush of self-certified doctors appear, prepared to prescribe remedies of the most fantastic kind, almost always including an element of shifting some major responsibility for electric power supply to government. These prescriptions are seldom the result of any process of systematic or thorough analysis. Most frequently they come from some, not very deep, knowledge of what is being done elsewhere in the world. "Why," the doctors ask, "shouldn't we handle electric power supply in this country as it is handled in England, in France, in Italy, and in Russia? These countries have shifted from private enterprise to governmental operation. Does not our own recent experience indicate the inadequacy of private enterprise to handle such a critical function as electric power supply? Shouldn't we try to avoid future similarly traumatic experiences by creating a modern national grid network operated by the government?"

Indeed, the Nixon administration would not surprise me if it came forth with a misguided proposal to solve the whole problem by setting up a vast federal bureau to make long-distance projections of need, to develop plans, to carry out research, and to sponsor a huge program of building vast electric energy generating centers and a national transmission grid to give the country the supposedly ultimate in solidly established mechanisms for supplying this indispensable service of electric energy. If such a program should be carried out, it would quite likely end in catastrophe. Such a proposal was made and actively pursued by the Federal Power Commission in the late 1930s under a very dynamic Democratic administration. Several members of that Commission prided themselves on their "progressive" thinking in electric power, and they concluded that what the country needed was a 287,000-volt grid or network covering the entire central and eastern part of the United States roughly north of the Mason-Dixon line.

This particular project, fortunately, was submitted to complete technical-economic critical analysis, and it was not too difficult to demonstrate its complete inability to meet the tests of practical and economic performance. (This analysis was made by an industry group under the direction of John C. Parker, then a former chief executive of Brooklyn Edison Company, and myself.) There is no reason, however, to believe that the generation of such ideas could long remain the monopoly of one particular party, particularly when the concept of doom can be kept alive by such judgment as the following of Dr. Frank Fraser Darling:[2]

> But it seemed to us that the very large catastrophe to which we were heading would not happen as the result of any one cause, but as a culmination of several factors, famine, war, pestilence, pollution and so on. A breakdown in technology, such as the New York–New England blackout of November 1965, could be the precipitating factor in a society geared to the electronic control of so much of the machinery of living.

Thus we have the spectacle that when a private system organization is put to a severe test and fails (as it did in New England in 1965), instead of making deep and competent inquiry, the public hears from all kinds of doctors and considers serious major changes. It is suggested that (a) the old organization, having suffered a failure, must continue to fail, even if failure is not necessary or inherent in the organization; and (b) since the public has lost confidence in what exists, concern for the future is impossible under private enterprise. In consequence, any effort even by earnest men of good will to determine a proper course—the best from the standpoint of the public interest—is made particularly difficult because of the dogmatic judgments arrived at by

[2] Frank Fraser Darling, *The 1969 Reith Lectures* (London: British Broadcasting Corporation, 1970).

large numbers of people without first determining where fact and truth lie.

One important element of confusion can be traced to the tendency to regard private power with suspicion and to associate a kind of idealism and economic euphoria with public power. This notion, which springs from a memory of conditions long past, is reinforced by the knowledge that the U.S. organization of electric power runs counter to that prevailing in such economically advanced countries as England, France, Italy, Mexico, Sweden, and of course Russia, which have either completely or partially nationalized their power systems. On the other hand, Americans traditionally have a healthy pragmatic skepticism about encroachments of government; most Americans are rightly fearful of bureaucracy and an officialdom that manage often to combine undue authority with a sense of mission. Hence, it is no easy task to consider the question of public and private power: whether and to what extent it is necessary or right that government projects in this field should be allowed to expand and proliferate, and how existing projects can best be controlled and regulated in the public interest.[3]

Therefore in this, the first lecture, having traced the development of public power to its present strong minority position (23%) in this country over a period of half a century, I want at least to indicate the organization of electric power in half a dozen of the world's principal economically advanced societies and then try

[3] For a more detailed discussion, see Philip Sporn, *Vistas in Electric Power,* vol. 1 (New York: Pergamon Press, 1968).

to discuss the philosophical or ideological basis for publicly organized electric power. In my second lecture, I plan to examine the technical competence and performance under the two sets of organizations. Then in the third lecture, I plan to analyze the performance of power systems under different social organizational setups in protecting the national interest under normal and emergency situations. Finally in the last lecture, I plan to examine power system performance not on the basis of previous problems the power systems had to meet but on the basis of future problems.

T

THE ORGANIZATION OF ELECTRIC POWER SUPPLY IN ENGLAND, FRANCE, MEXICO, RUSSIA, ITALY, AND SWEDEN

E

England

In England and Wales service is provided by 14 statutory bodies: the 12 area electricity boards, the retailers of electricity; the Central Electricity Generating Board, the producers and wholesalers of electricity; and the Electricity Council, the central council of the supply industry.[4]

The area boards provide the service to consumers, who at the end of 1968–1969 numbered 18 million. In

[4] The Electricity Council Reference Section, *Reference Papers, Serial No. R.P. 1* (London: The Electricity Council, December 1969).

that year they required a supply of over 160,000 million kwh.

The English supply system, now a completely publicly owned system, is an evolution of a long line of development in electric supply that, being biased toward municipal ownership, resulted therefore in limited areas of service, discouragement of transmission, and discouragement of private enterprise by the impractical reversionary purchase clause in the 1882 Electricity Act. These impediments led to a mushrooming of numerous small undertakings, some direct current and some alternating current, of numerous frequencies, voltages, and practices. The ill effects of this system compelled reorganization.

After a number of ill-fated attempts, the Electricity Supply Act of 1926 was enacted establishing the Central Electricity Board charged with the responsibility of building a national grid over which electric energy from selected generating stations could be supplied wholesale to local distributors, who were left undisturbed. The grid was first placed into operation at the end of 1931 and resulted in the beginning of nationalization of centralized generation and coordinated operation. However, it hardly had finished its shakedown period when England and Europe were shaken by World War II. Although the Central Electricity Board helped the country through the rough 1939–1945 period, it was the Labor government's return to power in 1945 that triggered the complete nationalization.

The Electricity Act of 1947 brought the power supply industry under public ownership, and the existing operating groups, which then numbered 560, were integrated into new statutory Electricity Boards appointed by the minister of fuel and power. These boards, 12 in England and Wales and two in Southern Scotland (Scotland later came to be controlled by a separate power authority), shared responsibility for the retail distribution of electricity to consumers. Generation and main transmission, together with central coordination and policy direction, were the responsibility of the British Electricity Authority.

In early 1968 the minister of power began an examination of the organization of the electricity supply industry, and in mid-1969 he announced broad proposals for a reorganization. These were based on his conclusion that the central direction of the industry needed strengthening and that the existing degree of autonomy of the Central Electricity Generating Board and area boards was not acceptable. The minister concluded his statement with the hope that, by the reorganization he proposed, the advantages of central direction of major policy, then lacking, would be obtained. At the same time the advantages in terms of managerial responsibility and local initiative, the features of the existing system, would be retained in larger autonomous boards.

Of course this is an almost universal problem: how to achieve the benefits of size and centralization without losing the advantages of intimacy of contact between the operation and its consumers as well as the sympathetic response to human needs on the part of the operation.

The net effect is that during the last 20 years the power system of England has, as a result of unification and growth, given a good (but not particularly superior) performance.

F

France

Before World War II, the generation and distribution of electricity in France were entirely private, except for a small proportion of municipal distributions and for the Compagnie Nationale du Rhône, which was a mixed public and private company.

Generation was about half steam and half hydro, the hydro being located in the southern part of the country. This was favorable to an early development of interconnection. Before the war the transmission lines of the country were already meshed in a sort of grid, and the synchronous operation of the various systems was becoming general, although the 220-kv ties were still weak.

During the high-water periods (resulting from rain and melting snow in the Alps), the lines were not capable of taking all the available hydropower, and much energy was lost over the spillways, a situation that stronger transmission lines could have averted.

Apart from the general political-ideological drive for nationalization following the end of World War II, the French decision to nationalize the power industry in April 1946 was influenced by a number of additional factors relating to the three following major fields, and thus Electricité de France (EDF) was born.

Generation. An impression that with a nationalized or-

ganization less water would be lost over the spillways.
Distribution. The excessive number of small companies, the considerable variety of tariffs, and the impression that the great number of medium voltages did not correspond to an economic organization.
Construction. The fear of difficulties in financing the very heavy pending expenditures to reestablish a system that had suffered from the war and had to be expanded very quickly so as to be again capable of coping with the fast-growing demand. The public was frankly tired of rationing and blackouts.

The areas in which substantial changes came, probably more quickly than they could have with the preceding organizations, were chiefly in the areas of tariff, distribution and transmission, and cost.

The tariffs were made more rational which erased the impression that they were arbitrary; the principle of tariffication according to the marginal cost of development met with public favor. While prices of electricity varied with the region, voltage, and the size of the block of power, they eliminated differences between similar consumers serviced by different companies inside the same region.

A decision was quickly made to standardize distribution and transmission at nine voltage levels, but this slight progress still included far too many voltages. However, voltages were to be changed as soon as possible without incurring an unjustified expense, and all the new lines and substations built were designed to prepare for a further decrease in the number of voltages.

These changes resulted in savings in the cost of apparatus by centralized purchasing, they improved the competitiveness of French manufacturers for exporting equipment, and they also established the impression with the public that the technical choices were being made rationally.

During the 25 years following nationalization, the development of the electric load has been very regular, with growth running at about the same rate as in the United States. Rationing and systematic regular shedding of load came to an end in 1949. In 1946 the total kilowatt hours consumed in France were 23 billion (in contrast to 22 billion in 1939 and the wartime minimum of 16 billion in 1942). In 1969 the consumption amounted to 130.7 billion.

Another important change during these 25 years has been a progressive decline in the percentage of hydro capacity built in the total generating resources. This decline has been a result of the progressive deterioration in the quality of the remaining underdeveloped hydro sites and the poorer economics of hydro construction and hydro generation as compared with steam.

Still another change was the marked increase in the size of steam-electric units. In 1939 the largest sizes were 50 and 60 mw. After the war the size jumped to 125 mw. Later a new standard was set by EDF at 250 mw and more recently at 600 mw. But on the whole it would appear that this has been much slower than it might have been under the previous organization.

A nuclear program was started as soon as the studies

of the Commissariat a l'Energie Atomique had completed the experimental work involved in building a natural uranium unit. France's technical bet was initially placed on the gas-cooled, graphite-moderated reactor, and for this purpose the French chose natural uranium. They switched over to enriched uranium at the Chooz station (located in France), a joint 525-mw unit in partnership with a Belgian group, and they are now in process of installing, again in partnership with a Belgian group, a Westinghouse pressurized-water reactor at Tihange, located in Belgium.[5]

However, France has for a long time been in technologic-economic atomic doldrums, mainly because of the strong intervention of General de Gaulle. This remained unabated throughout his lifetime; the expected changes following his death have just appeared on the horizon.

M

Mexico

The development of electric power in Mexico was started by private capital, much of it foreign, and continued during the first half of this century. The performance under these auspices was fair. There was some pioneering: The first hydroelectric plant in this hemisphere was constructed at Nacaxa on the Gulf of Mexico at the turn of the century. A transmission line was built to Mexico City to bring this power to market. With

[5] Philip Sporn, "World Nuclear Energy," *Columbia Journal of World Business,* vol. 4, no. 5, pp. 71–80, September–October 1969.

growth in population and increasing industrialization, the private companies' performance began to display weakness—mainly lack of capital resulting from inadequate earnings. Thus, during the period 1938–1951, energy growth went from roughly 2512 million kwh to 4900 million kwh, or a compound growth of 5.5%.

The revolution of 1910 resulted in the present democratic government of Mexico. The society that came about has now evolved into a private enterprise economy with a strong mixture of socialism.

Speaking in Chicago on July 16, 1961, on his government's policy in regard to outside investment in Mexico, the minister of industry and commerce said:

> The State has limited its activity to the development of the economic structure: highways, electric power plants, schools, hospitals, and the development of public services such as railroads, petroleum and electricity. Only those public service enterprises that the Government considers fundamental to the welfare of the majority of the inhabitants are in the hands of the nation.

The state moved into the vacuum caused by an inadequate electric energy supply by the privately organized industry. Thus from 1946, the installed capacity grew more rapidly with increasing participation of the Federal Electricity Commission (established in 1937) and its affiliated companies. By 1959 the condition had completely changed from what had existed in 1938. Whereas capacity did not quite double in the 13 years between 1938 and 1951, it increased by 161% in the ten years from 1949 to 1959—from 1,097,000 kw to 2,739,000 kw. Energy production between 1938 and 1951 increased at the rate of only 5.5%, while between 1949 and 1959 it increased at the rate of 8%. More recently, the energy production between 1960 and 1967 increased at a compound rate of 10%, but more and more of this was generated by state companies.

The current policy of the government is that the generation, transmission, and distribution of public electric power is reserved solely to the nation and no new concession can be granted to private individuals or companies. Operating under this policy, approximately 84% of electric generating capacity was, by 1966, represented by facilities of public agencies, including the acquired private companies. The remainder was contributed mostly by captive industrial plants, and these were privately owned. And while nationalization improved the Mexican power situation enormously, it still did not give the Mexican people, as will be shown later, a satisfactory power supply system, even though the government has actively intervened in this area for the last 20 years. The criterion by which this judgment is reached will also be discussed later.

R

Russia

Electric power in Russia is nationalized, as is all industry. But electric power occupies a special place in Soviet economic society as well as in Soviet ideology. There are a number of reasons for this. (1) Lenin had early identified the victory of communism with electrification. (2) Communist bureaucrats and economists use electricity production as a key factor in "the construction of the scientific-technologic base" of communist society. (3) Since the beginning of the commercial production of electric energy in 1882, the United States has been the world's largest producer and currently accounts for about 36%. Russia is a very bad second with about 15%. For many years it was both the announced and actively promoted objective of those charged with the development of the Russian economy to reach a position of equaling and surpassing the United States in electric energy production by 1980.

Twenty years ago Soviet electric energy generation was increasing at a compound rate of over 13%. In 1969 the culmination of a long series of declines in the growth rate was such that the Russian increase in production in that year was less than the American increase: Russia's production was 689 billion kwh, an increase of 7.8%; United States production was 1552 billion kwh, an increase of 8%. For the first six months of 1970 the Russian increase was 6% but that of the United States was 6.8%.

In a previous analysis of a report of the USSR Central Statistical Administration on "Plan Fulfillment for 1968," I stated:[6]

Thus we have in this short period seen the Russian figures (on electric energy generated) decline from 13% to 9.5% with a clear indication that this decline will continue. Further, there is considerable basis for judgment that the time is not far off when the Russian increase in

[6] Philip Sporn, *Electrical World,* 25 November 1968. The USSR lags far behind in the power race with the United States.

energy generation will not only be significantly below 9.5%, but will drop first to the worldwide historical trend line of the power industry 7.2% and thereafter probably below this figure.
Since 1969 we have seen the continuation of this drop toward the worldwide figure of 7.2%.

The figures of generating capacity commissioned in 1969 are also of interest: The Russian figure is 12,000 mw and the United States is 24,018 mw, more than double the Russian capacity. The Russian estimate of capacity to be installed by the end of 1970 is 180,000 mw, less than half of the United States estimate of 364,009 mw. There is a strong probability that the Russian figure will be even less than 180,000.

Thus, if we consider Lenin's dictum, that "Communism is Soviet power plus electrification of the whole country," and Russia's certain inability to surpass the United States in electric energy generation by 1980 as an index of Russia's failure to achieve its other economic goals, we may have in this electrical confrontation as solid a basis as one could desire for the belief that, at the rate Russia is now moving toward surpassing the American economy in any foreseeable future, the USSR will find it a most formidable task indeed. It certainly will not achieve it by 1980.

I

Italy

Electric power became nationalized in Italy with the establishment of the National Electric Energy Agency

(ENEL) and the transfer of the electricity supply enterprises to the agency by a law promulgated on December 12, 1962.

This law reserved to ENEL the activities of generating, importing and exporting, transmitting, transforming, distributing, and selling electric power produced from any source except certain minor operations that had been given specific power to generate under previous laws.

ENEL is subject to the supervision of the minister of industry and commerce, conducts its activities according to the directives issued by a committee of ministers presided over by the prime minister or by a minister delegated by the prime minister, and composed of the ministers of the budget, treasury, industry and commerce, public works, state holdings, and agriculture and forestry. This becomes a very complicated administrative body. Nevertheless, the objective of the law (#1643, December 6, 1962) is very clearly stated in Article #1. In the general interest, the Agency shall provide for the co-ordinated utilization and development of the plants, with a view to ensure the supply of electric power at the lowest operating costs, adequate in quantity and price to meet the requirements of the balanced development of the national economy. Essentially these are similar to the directives under two laws in the United States, the Federal Power Act of 1935 and the Holding Company Act of 1935.

ENEL has been operating under this law for nine years. In its report and balance sheet of December 31,

1969, there is a good strain of *apologia pro vita sua.* Thus, in its balance sheet of December 31, 1969, the agency states:[7]

> It should also be taken into account that when the Agency was first constituted, only a single thermo-electric plant of over 160 MW was in service, while today there are eighteen such plants in service, seven of which are of 320 MW and two of 600 MW, and a further forty-seven units of 320 MW and six of 660 MW are at the planning or construction stages.

Again, the agency states:

> We have been extremely active in this field (strengthening of plant), too, and have used every means possible to obtain results of undoubted importance. One need only recall, for example, the reduction achieved in the average travelling distance of very high tension power of the transport network, the reduction in the consumption of fuel, the adoption of the use of high power in thermo-electric stations and the unification of plans, increasing automation, the unification of administrative and accounting activities, the quarterly collection of debts, and the adoption of new techniques and procedure.

In 1969 the agency generated 75,897 million kwh and sold 68,674 kwh. An interesting figure here is the loss of 7223 million kwh or over 9.5%, an extremely high figure for a country as geographically compact as Italy. In other respects, the job of electrifying the country is not as yet of a particularly high order.

One cannot help but wonder what brought about nationalization. At the time the law was passed, I had occasion to go into this question with several Italians, from both the private utility groups and the political

[7] *The New York Times,* 14 July 1970.

sector. I came to the conclusion that nationalization was the result of a complicated mixture of actions and interactions between a socialist-communist ideology, political maneuvering in a country that is politically among the most unstable on the entire European continent, foolish and arrogant behavior toward people in government on the part of private power utilities, and of unnecessary intervention in this most secular of activities by the Vatican itself.

S

Sweden

The Swedish power system[8] forms a block of 15,000 mw interconnected to a grid system covering the whole country. Of this, nearly 11,000 mw is hydropower, the rest consisting of oil-based back-pressure power and peak power. During a normal year hydropower covers more than 90% of the energy production.

The 1970 per capita consumption of electricity is among the highest in the world, amounting to about 7500 kwh. The demand is expected to increase by 7% to 8% yearly during the 1970s. A major contributing factor to this tendency is the recent rapid growth of electric space heating.

Industry is responsible for a major part of the consumption, about 60%; railways and communications

[8] I am indebted to my friend, Mr. Eric Grafstrom, president of the Swedish State Power Board, for this authentic and concise summary of the organization of electric power in Sweden. With some recasting and some minor editing, I have used his account in totality.

take about 5%; and the remaining 35% is used by retail consumers.

About 85% of the population of Sweden is located in the southern part of the country, and about 85% of the hydropower is located in the northern part. Thus transmission lines at very high voltages were required: As early as 1952 the first 400-kv line was commissioned. The Swedish trunk line system now comprises seven north-south lines at 400 kv and six 230-kv lines.

In Sweden about 45% of the annual power production is the responsibility of the state Power Board, about 40% is in private hands (the private power utilities of importance are about ten in number), and the remaining 15% is provided by municipalities. Power exchange between all utilities is arranged on a voluntary basis and is now very far advanced. In spite of the mixed system of state, municipal, and private companies, the technical and economical operation of the system works in the same way as if there were only one utility. The state has not influenced the structure of the production and distribution systems toward any special kind of organization or to create a monopoly. The only departure from the free market system is that the state has given the Power Board the exclusive right to build and own the trunk line system for voltages of 230 kv and above.

According to an agreement between the Power Board and the other power producers, the planning and the operation of this trunk line system is handled by the Power Board in consultation with the other contracting parties. The same agreement states that the contracting parties have the right to transmit their power over the line network. The transmission fees are based on bare costs, with the Power Board providing the capital without requiring any other return beyond normal interest and amortization.

Swedish electricity production has so far been dominated by hydropower. However, Swedish hydro resources are limited. In order to meet continually rising electricity requirements, other sources of power must increasingly be developed. During the 1970s it is expected that hydropower will supply 13% of the increase of capacity, oil condensing and oil back-pressure power 20%, nuclear power more than 40%, and the balance of about 25% distributed among different kinds of peak power.

Nuclear capacity will probably comprise about 22% of the total capacity in 1980. In energy production, nuclear capacity will probably account for one-third of the total, or 50 billion kwh, because of its capital cost characteristics.

In the cities, where it is combined with a district heating system for dwellings, oil-fueled power operating on a straight condensing cycle and oil-fueled power operating on a back-pressure cycle, for the last few years have come, and will continue for some years, to dominate the new production plants. The first commercial nuclear plant (450 mw) will be commissioned in early 1971. In 1973 and 1974, two big nuclear power units, each of about 800 mw, are expected to start operation, and from then on similar units are expected to be commissioned yearly.

All nuclear plants planned for the 1970s are of the light water type, either of the pressurized or the boiling water type. The work on a special Swedish heavy water type has been discontinued. Making such a heavy commitment in nuclear power quite obviously involves a rather drastic step into a new technique for the Swedish power industry. The lack of domestic sources of coal and oil and the resistance that a further exploitation of hydropower has been meeting as a result of the recent concern with environmental factors have contributed to this general attitude among the power producing people. At the same time, they believe that nuclear power will give the lowest price for electricity in Sweden in the years to come. Therefore, the power producing people are trying to safeguard the production system against the uncertainties and possible outages to be expected during the introduction of nuclear plants by increasing the capacity reserves to a higher percentage than will probably prove necessary in the long run.

The Swedish power supply system has integrated public and private power and existing and new technologies to a remarkably successful degree. This is equally true of the system's enlightened and pragmatic views on social costs to be allocated to power, whether public or private, so as not only to gain for the state important sources of revenue but to keep competition alive between the public and private sectors on the highest level of fairness and ethics.

T

THE IDEOLOGICAL CONSIDERATIONS INVOLVED IN THE MERITS OF PUBLIC VERSUS PRIVATE POWER

It is most pertinent to ask what differentiates public from private power. Surely no public power exists comparable to a public street or highway, and no private power exists comparable to a private estate or building. The business of electric power supply is a public one in that its service has to be made available to all the public within the service area of the system. Thus, ownership is the only principal difference between so-called public and private power—in one case ownership by government or government agencies, in the other ownership by private individuals or corporations. As a general rule, the privately owned electric power systems are subject to the jurisdiction of numerous regulatory bodies, in many cases of both state and federal bodies.

We thus have this paradox: Private power, as represented by investor-owned utilities, dedicated to public use, is under the regulation and supervision of a large number of publicly created bodies forming major parts of our state and national governments, with the object of protecting the public interest. On the other hand, in most cases, public power is subject to no comparable regulation.

As a consequence, the public has no protection on rates covering service rendered by public power bodies. This lack of regulation means not only that the public power entity is engaged in no tension-breeding relationships with a regulatory body but also that the public is deprived of any real opportunity to participate in decision making relating to rates, terms and conditions of service, and environmental protection. Such public involvement would be impossible in Russia and because of the difference in the political mores, may mean little in France or Mexico. It does, however, provide a meaningful means of public expression and, if properly channeled, for exerting public influence on a vital service in the United States.

When the city of New York recently found it necessary to raise additional money for budgetary purposes, it apparently was not too difficult for it to decide, primarily on the judgment of one man, the mayor, to increase water rates by 75%, even though they had been increased 100% less than two years previously and the new increase would thus mean a compound increase of 250% in less than two years.[9] It is hard to see how this arbitrary lack of concern for public interest or public morality is any improvement on the almost century-old philosophy of the public be damned, given utterance by Commodore Vanderbilt, circa 1883.

Another example is the concept of private profit, which is often used by proponents of public power in flagellating the private utilities. Without even remotely suggesting the attachment of a stigma, no matter how slight, to private profit, how private is profit in a public service enterprise when the participants in the form of equity ownership are insurance companies, foundations,

[9] *The New York Times,* editorial, "And Paying for Them" 11 June 1970.

college endowment funds, charitable organizations, hospitals, pension trust funds of innumerable institutions, or any individual who wants an outlet for his savings? How private is profit in that case?

Water power is one of the most interesting examples of an idea that has managed to accumulate an unimaginable number of layers of the hot lava of controversy but, in the net, invariably gains support for public power. Even though water power will eventually become an almost negligible item in the total primary energy resources contributing to our electric supply, it has had a history, now pretty well ended, as an important means to an end—a means of producing more economical electric energy. That simple idea of water power producing economical energy has become befogged by the concept of water power as a natural resource, almost sacred, to be kept for development by the people. Hence according to this philosophy, when you develop water power, it follows that it should be a public power development. Since the water itself is apparently free, it is quite easy to make the average person believe that water power results in cheap electric power. Thus an almost sacred association: water power equals public power equals cheap power. That water is merely a resource, which may or may not produce economical power when developed and exploited by an input of large increments of capital and labor, is almost completely ignored.

One of the strongest ideological arguments for public power is costs; public power, supposedly, gives lower costs to the consumer. But does it really? There is solid reason for doubt. I shall go into this aspect much more deeply in Lecture 3. Here I shall only note that it is ironic that the very marked characteristic of widely scattered individual decision making that is so cherished a part of the private enterprise philosophy and that carries with it so high a price in social costs (such as technological unemployment, pollution, deterioration of cities and particularly centers of cities) is taken advantage of and exploited by the managers of socialized industry. Thus freedom from local, state, and federal taxes, which results in inadequate support of policing, education, health, and even defense activities, is justified on the grounds that it makes no sense to take money out of one pocket and to put it into the other. In truth, this is but a glib and superficial rationalization of a bad socioeconomic practice that results in taking money out of the pockets of a large segment, in some cases the great majority, of our population and distributing it to a relatively favored few—and those not even selected on the basis of any special merit or need but on the basis of political expediency or usefulness in continuing a particular group in power.

The damaging effects penetrate even deeper. Out of a mistaken disregard of the cost of taxes—costs that provide funds for defense, education, research, pollution amelioration—there too easily develops the notion that money is something that any sovereign government can make available at low cost, at no cost whatsoever, or even at negative cost, without the transfer of heavy burdens to other segments of that society not involved in the allocation and that this can be done without limit.

The hard fact is ignored that money utilized constitutes a demand upon or an allocation of a society's finite or limited resources: materials, research, fabrication plants, and human labor—unskilled, semiskilled, and highly skilled.

This makes it altogether too easy to misjudge completely the socioeconomic desirability of projects, many of them of a major scale, in comparing the costs of several alternatives. Instead of having costs based on actual demands upon real resources, they are based on distortions resulting from artificially contrived monetary arrangements.

Advocates of public power contend that only when power is public can one look forward to adequate and effective planning and development. If the relationship is examined between gross national product and electric energy production in the United States, the following is observed: Between 1950 and 1968 the gross national product went up from $284.8 billion to $865.7 billion, an increase of 205%. In the same interval, electric power production went up from 329.1 billion kwh to 1327 billion kwh, an increase of 305%. The gross national product is expressed in terms of current dollars of declining value, whereas the kilowatt hour has not been altered at all in its energy value. The result of this development is that, by the end of 1967, the per capita consumption of electric energy in the United States was the highest of any industrialized country with the exception of Sweden and Canada; the latter countries are unique from the standpoint of their very large hydroelectric resources and their relatively small population.[10] Per capita use in Sweden and Canada—each partly nationalized—was 6762 kwh and 8111 kwh, respectively. Corresponding consumption in the United States in 1967 was 6612 kwh. This compares with less than 3800 kwh for the United Kingdom, less than 2300 kwh for France; in both countries electric power is nationalized—and with slightly under 2500 kwh for Russia—all three completely nationalized. These figures offer no support whatever for the argument that only public power guarantees adequate planning and development.

Another argument frequently used for public power is that that is the surest method of protecting the public interest. This is not only open to question but there is good reason for believing that the reverse is more likely to be the case. One of the first things that happens when any given undertaking becomes public is the development of an immediate fuzziness of thinking that automatically merges the interests of the state and the public. However, government, the state, does not always act in the public interest. More than 20 years ago *The Economist* (London)[11] pointed out:

> It is far nearer the truth to say that "the state" merely consists of a number of government departments, each of which tends to put its own interest above those of the public. No one would dream of asserting that the state's interest and the public interest are identical where the civil and political rights of the individual citizen are concerned. Indeed, the history of freedom is one long list

[10] Fremont Felix, "Electric Energy Annual Growth Rate on Downward Trend," *Electrical World,* 6 July 1970.
[11] *The Economist* (London), lead editorial, "State and Public," 18 September 1948.

of devices—from trial by jury and habeas corpus to the judges' rules of evidence—based on the assumption that the state cannot be trusted to serve the public interest. . . . Yet in economic matters the doctrine appears to be accepted that whatever a Minister of the Crown may take it into his head to do is ipso facto in the public interest.

Conditions have not changed much in the last 20 years. Thus in 1963 the then chairman of the Federal Power Commission (FPC), Mr. Joseph Swidler, testifying before a committee of the House of Representatives about the relative roles of the private power industry and FPC, was engaged in the following colloquy, as it appeared in an official transcript of the hearing, on the motives of the private power industry:

Mr. Swidler: Mr. Chairman, these are regulated industries. They make the same profit whether they do a good or poor job. What do they care? They still get their 6 or 7 per cent.
Mr. Thomas: Yes, but they will make more profit if they tie in, as you pointed out.
Mr. Swidler: No sir, they do not. They make the same profit. It is a matter of indifference to them whether they do a good or a bad job from the point of view of the national interest. The people who care . . .
Mr. Jonas: I would like to have the record show . . .
Mr. Swidler: The people who care are the Congress and the Federal Power Commission.

When some days later I gently chided Mr. Swidler for making such an outrageous statement in the light of his personal acquaintance with a number of private power people whose record of performance was a flat contradiction of his statement, he said he misspoke. But there is a substantial number among our 204 million people who have this same unbounded faith in a supposed universal and ubiquitous coincidence between public employment and concern for the public interest.

It is frequently stated that because electric power has developed at such a rate and the size to which projects can grow are so stupendous, if a country's supplies were in private hands, many projects would fail of execution for lack of private resources. This had some plausibility perhaps in the atmosphere of economic paralysis that afflicted the country in the early 1930s, but the situation today is quite different. Today the utility systems of the nation are adding new capacity roughly at a rate of 20 million kw per year, and it is hard to imagine that a single national project in power would be commensurate. Or if the measure of an undertaking is dollars, this argument becomes flawed in the face of the $440 million project of Ohio Valley Electric Corporation (OVEC), the development of which was a challenge 18 years ago to the private power industry of the Ohio Valley. The industry accepted the challenge and proceeded in the remarkably short time of less than three years to bring into being a new and complete power system capable of furnishing the AEC Portsmouth operation with 25% more electric energy per year than in use at that time by New York City.[12]

Today a project triple that size would seem to be an entirely feasible undertaking by several individual members of the private utility industry and could certainly be brought into successful completion by a com-

[12] Philip Sporn and V. M. Marquis, "The OVEC Project," AIEE Winter General Meeting, New York, January 1954.

bination of a number of them, even though the initial commitment were to carry with it the further responsibility of trebling within a period of another 15 years.

Recently the expansion of electric generating capacity has been stopped or delayed at a growing number of points in the United States. These delays are a result of well-intentioned activities that have caused rising public anxiety about the environmental impact of the operation of electric generating stations, and particularly of atomic generating plants.

This concern over environmental effects has now reached a point where those charged with the responsibility of maintaining the needed growth in the energy supply find themselves unable to carry on effectively. Progress in developing hydraulic energy sources is stalled as a result of concern for the landscape; management is inhibited from constructing fossil-fuel generating plants owing to considerations of atmospheric pollution; and in some localities cannot construct, or operate after construction, atomic energy plants because of concern about the release of radioactive substances to the atmosphere and to water. Another cause for delay is objection to the thermal pollution of the water bodies utilized to condense the exhaust steam. Because of delays in the installation of new generating capacity, many major power grids are without comfortable reserves to meet emergencies. If this opposition to expanding the electric energy supply continues, surely close to a catastrophic situation will result.

One would expect that this concern for environmental protection, which in many cases might seem difficult to reconcile with an adequate treatment of the problem of profitability of an enterprise, would lead public enterprise to assume a position of leadership in showing the way of how to solve such difficult environmental problems. However, in electric power the leadership in developing technology to solve the problem of air pollution or thermal pollution came not from the public but from the private enterprise sector.[13, 14]

As long ago as 1952 in the case of air pollution, 682-foot stacks were built at the Clifty Creek plant of OVEC. In the 1950s in the case of thermal pollution, cooling towers for the full condensing requirements of the 675-mw Clinch River plant of Appalachian Power Company, located on the extremely low flow Clinch River, were developed. More recently, hyperbolic cooling towers were used for the 1600-mw Mitchell plant of Ohio Power Company operating on the Ohio River to avoid direct use of that major stream.

After integrating all the ideological arguments for public power, it is clear that they do not make a very convincing case. Yet, the fact that we have traveled as far as we have along the road of public power indicates the strength of the naïve faith in public power. The strength of that faith warrants respect and concern.

For example, *The New York Times* is a completely unregulated private enterprise operation. Its editors certainly are aware of the fact that our economy is, in the

[13] Philip Sporn and T. T. Frankenberg, "Pioneering Experience with High Stacks on the OVEC and AEP Systems," presented at the International Clean Air Congress, London October 4–7, 1966.

[14] H. C. Barnes, T. T. Frankenberg, and C. P. Lugrin, "Hyperbolic Cooling Tower Highlights the Features of AEP System's Big Sandy Plant," *Power Engineering*, December 1962.

main, a private enterprise one. Yet in commenting editorially in August 1953 on the effect of the change in political party control of the White House on a possible shift in policy regarding electric power, *The New York Times* said:[15]

On the other hand, there is not necessarily an overwhelming case for federal action when it comes to projects for the production of electric power that can be carried out locally under public regulation.

I commented to the editor that it seemed to me that this statement deserved a prize as an outstanding example of modification, delimitation, and almost complete sterilization of a statement of principle.

Disregarding the fact that the editor never saw fit to make a reply, if it is not clear that there is no case for federal action when it comes to projects for production of electric power that can be carried out locally under public regulation, is there any case where *The New York Times* believes that federal action is unwarranted?

There must be some solid reasons for public power. The fact that public power accounts for 23% of our electric power supply is in itself warrant for that conclusion, but *surely it is not a conclusion based on ideology*. Therefore, to find a basis for it we will have to explore further. This I hope to do in the succeeding three lectures.

[15] Editorial, "Policy on Power," 21 August 1953.

2

Lecture Two

TECHNICAL COMPETENCE AND PERFORMANCE AS INFLUENCED BY SOCIAL ORGANIZATION

There is no question that the business of supplying electric energy to a modern society is technologically highly involved: Its development is capital intensive to a high degree and it is risk laden, economically. Therefore, the economic success for both producer and consumer is bound to be heavily influenced by differences in technical competence and performance. It is necessary to examine whether these differences stem from and are influenced by the differences in social organization. To this end, it is well to examine as far as we can the relative performance of public and private enterprise operations in electric power.

T

TECHNICAL COMPETENCE

The truism that in the final analysis the success of enterprise is dependent upon people is as solid for governmental enterprise as it is for corporate enterprise. The source of people in either case is the population of the particular society. Offhand one would expect that the kind of people you attract to either public or private enterprise would be the same and the results they obtain would also be equal.

While this is a good theoretical concept, it is doubtful whether it is true. Performance ultimately depends not only on people but also on organization. If the same kinds of people and the same organization run both public and private enterprise operations, they should be able to show the same performance. In the final analysis, a consideration of the relative quality both of the people and of the two types of organization is basic to the concept.

To shed light on this point, we now have a mass of information and experience in the United States on the staffing and performance of a large number of governmentally administered or quasi-administrative, quasi-legislative agencies. A group of agencies, such as the Interstate Commerce Commission (ICC), the Federal Power Commission, Securities and Exchange Commission, Atomic Energy Commission, Bonneville Power Administration, Rural Electrification Administration, and National Aeronautics and Space Administration, all can shed light on this question, why there undoubtedly eventually develops a divergence in quality between people in public and private enterprise.

It is very interesting to pursue further the experience of anyone of the number of public agencies enumerated. Almost without exception, they start with an aura of high public esteem and great public expectations. As a consequence, they all tend to attract capable, bright, ambitious young people with a strong motivation to make a contribution in the public service area and in the public interest. They invariably start in with a staff that has an exceptionally large percentage of good and even very good men.

But this is merely an initial or sort of honeymoon phase. It is followed by a diminution of enthusiasm for the work of the organization and the introduction into its staff of people who quite legitimately seek employment as a foundation or stepping-stone in the building of their careers. They never intend to make the particular organization the locus of their lifework. Rather, they want to gain some experience, to develop governmental contacts—possibly in a new area of activity—and after a limited period, to use this experience as the basis for finding employment in other areas.

As these people leave, they are replaced by others of similar or possibly of less quality. In the final equilibrium stage, the entire staff consists of a rather mediocre group of politicians, career people, and lower echelon white-collar workers; almost all lack great ideals, ideas, ambition, initiative, or competence and possess little concern for much beyond their pay and their eventual retirement. An excellent case in point is the ICC in its most recent 25 years of existence.

A good example of a grand honeymoon phase exists in the current organization and operation of the Tennessee Valley Authority (TVA). The overall high standard of performance to date in that governmental operation is because of the high quality of the first directorship personnel, such as David Lilienthal, Gordon Clapp, G. O. Wessenauer, and a number of others. It is hoped that this quality would continue, but already there is some evidence to warrant the view that great optimism is needed to believe that it will.

The tenure of office of many of the people, certainly those with important responsibilities in such organizations, has a degree of impermanence about it even at nonpolitical levels. Sooner or later, with a change in administration, hints are thrown out that "we would like to have a man sympathetic to our program in the position you occupy." While a stand can be taken against such a not-too-gentle hint that he ought to resign, the higher the quality of the man, the less inclined he is to fight such pressures. This political disruption literally raises havoc with the continuity of plans that are highly involved and of great moment to the future of the organization.

While the pay for many technological positions in government service in almost any discipline—engineering, accounting, law, economics—may be higher than it was only a short time ago, it is still inadequate for those persons of quality who serve with great dedication and where service of the same quality is compensated by substantially higher rewards in private enterprise.

Where does this lead to? It leads to restrictions on the productive development of men and systems in public operations. The reasons for this are numerous: (1) to begin with, complete identification with any complex project that is bound to extend over a considerable period of time is almost impossible because of the lack of continuity of service, if not for the individual then for his supervisor; (2) the restrictive limitations on authority; and, finally, (3) the inability to cross lines of authority except at great peril to the individual.

This case is not unique to the United States. It is true in almost every democratic society. In England, for example, one of the great scholarly and highly dedicated public servants, Sir Ronald Edwards, the past chairman of The Electricity Council, in a lecture delivered some three years ago quoted Lord Beeching:[1]

> In the case of private industry there is a single, clear and unchanging primary objective, and management is free to optimise its behaviour in pursuit of that objective within a general framework of conditions imposed by law, practice, and opinion, which it hopes and expects will also be reasonably stable. Therefore, its actions can and should be compatible and coherent.
>
> In the case of a nationalized industry, on the other hand, objectives are more numerous, more ambiguous and less distinguishable from qualifying conditions. Moreover, they fluctuate in their supposed order of priority, not merely from Government to Government, nor even just from year to year, but almost from day to day

[1] Sir Ronald Edwards, "The Stamp Memorial Lecture, November 1967," University of London (London: The Athlone Press, 1967).

at the whim of public and parliamentary opinion. Imposed policy decisions which are bound to increase losses or limit capacity may be followed within days by a howl of anguish about the financial state of the organization, or about shortage of supply. Management does not have the freedom to optimise its own performance in pursuit of a single objective, or even in pursuit of a number of stable and compatible ones.

The peril of an unsuccessful exploitation of an idea is much greater to an individual in public than in private enterprise. Any idea or project of daring and imagination has in it certain elements of risk that sometimes cannot be seen, no matter how perceptive or analytical the person sponsoring it may be. In private enterprise, men have carried out projects of this type, failed, and yet have not ruined their careers. In public enterprise, the danger to the career of a person of even a single incident of this kind is much greater. Yet it is only in these areas where risk has to be taken that the great potentials for high-yielding success exist. Research and pioneering always carry with them an element of risk, a sort of racetrack gamble. The willingness to sponsor such risky projects, the willingness to expose oneself and one's career, are going to be determined first by the confidence of the individual in his own ideas, and then by his evaluation of the relative yields to himself, positive if he is successful and negative if he fails.

In a way this danger is not entirely absent in private enterprise. Here, too, the larger the organization gets and the shakier the tenure of those in high office, the less likely will there be a willingness to undertake risky, imaginative projects carrying with them potentials for great success or failure. In yielding to fear, the possibility of developing such projects is jeopardized. But the jeopardy is greater in public than in private enterprise.

With this as background, let us take a brief look at the performance of a number of outstanding public power organizations.

Consider, for example, the case of TVA. It serves a region of some 201 counties in the states of Kentucky, Tennessee, Virginia, North Carolina, Georgia, and Alabama. In 1969 it sold some 86.4 billion kwh to over 2 million customers of its own and of its local distributors. These sales constituted 6.6% of the total sales of the electric utility industry of the country. Interestingly, of its total generation of 90.59 billion kwh in that year, its hydro generation was only 14.99 billion kwh, or 16.6% of the total. Thus TVA obviously is primarily a steam-electric generating utility, by far the largest in the country.

The growth pattern of TVA sales is extremely interesting. It is shown in Table 2.1. This is a very fascinating table because, while it shows TVA progressing at a compound growth rate of almost 20% in the years 1950–1954, which had stepped up to a rate in excess of 25% in the years 1952–1955, it dropped to a rate of 17.5% in 1954–1957 to 4.5% from 1956 to 1961, and a bare 4.8% from 1958 to the projected 1973. The last percentage is based on the author's assumption of restoration of approximately 20 billion kwh in gaseous diffusion plant load. The explanation of this is obviously the great expansion in TVA's load that developed be-

Table 2.1 Growth Patterns on TVA Sales

Interval		Kilowatt hours	Years for Doubling	Growth Rate (%)
1950–1954	54	30.06 × 10^9		
	50	14.17	3.72	19.15
1952–1955	55	42.05		
	52	20.18	2.84	25.3
1954–1957	57	57.04		
	54	30.06	4.1	17.5
1956–1961	61	60.10		
	56	53.85	16.0	4.5
1958–1973	73	113.5*		
	58	56.72	15.0	4.8

Source: Annual Reports of TVA, 1965–1969.
* Assumed restoration of 20 × 10^9 diffusion load by 1973.

tween 1950 and 1957 as a result of the expansion of the gaseous diffusion facilities of the Atomic Energy Commission (AEC) at Oak Ridge and the new load at Paducah.

The original assignment to TVA of the responsibility for serving the gaseous diffusion plants was the result of an assumption that TVA was the only agency that could supply that load. It was later proved conclusively in the case of Electric Energy, Inc., and particularly in the case of the Ohio Valley Electric Corporation (OVEC), as will be shown, that this was not so. When the AEC was confronted with the need to contract its production of enriched uranium, this caused a slowdown in TVA's apparent rate of growth, even though the normal regional growth continued without interruption.

In the discussion of the ethical factors that should be considered in evaluating the relative merits to any society of public and private power, it invariably is pointed out that private power has in it an element of profit, and therefore, it presumably represents a higher cost. I shall deal with this subject of costs in detail in my next lecture. But here I want again to call attention to the fact that private power is under the regulation and supervision of a large number of publicly created bodies with the object of protecting the public interest. Public power is in most cases subject to no comparable regulation.

We had a very remarkable demonstration of this point in the recent history of TVA. In July of 1970, TVA announced an increase of about 25% in its power rates to be effective the following October 1, a percentage that would lift its yearly revenue by $115 million. *The Wall Street Journal*'s first dispatch that carried this item[2] pointed out that TVA had had two general rate increases in recent years, one of 7% in August 1967 and one equal to about $30 million a year in March 1969. Three days later, *The Wall Street Journal* carried another dispatch from Knoxville stating that TVA had confirmed its plans to effect a rate increase on October 1 that would boost its power revenue by $115 million and stated that the cost to the region's consumers would go up about 23%. The effect on the residential rate would be to increase the average rate per kilowatt-hour to about 1.25¢ from just over 1¢ in 1969. In contrast, the national average is 2.1¢. The dispatch blamed the

[2] *The Wall Street Journal,* 17 July 1970 and 20 July 1970.

increase on "rapidly rising costs for the coal burned in TVA steam-electric plants" and estimated the system total fuel costs in the year that started July 1 will "be about $60 million higher" than in fiscal 1970. This raises two points that have nothing to do with the question of the validity of this increase. It may very well be, and indeed it probably is, a fact that the entire increase is fully justified and indeed is necessary to TVA's financial health. However, the first important point from the public standpoint is that no one representing the public at federal or state level has had an opportunity to question TVA about its rising costs or to examine its quantitative analysis in regard to the $115 million increase and why it is necessary in view of the increases made in August 1967 and 1969. No private enterprise electric operation in the United States would be permitted to put in a rate increase without full justification, frequently involving hearings running into months; it is a very rare case where the full amount requested is granted by the regulating commission.

The second important point is TVA's explanation of the need for the increase in rates—the fact of "rapidly rising costs for the coal burned in TVA steam-electric plants" that will bring about system total fuel costs in the year that started July 1 "about $60 million higher" than in fiscal 1970. One cannot help but wonder where TVA's prestigious management skill, foresight, and ingenuity were while its fuel costs were developing a rise of close to 50% in one 12-month period.

Or consider the case of the Bonneville Power Administration (BPA). The Bonneville power system growth in the last 9 years has not been particularly unusual. Its revenues in the 9-year interval, fiscal years 1960–1969, increased by 79%.[3] This compares with the figure for the total electric utility industry of 78%.[4] And yet we find Mr. David S. Black, Undersecretary of the Interior, in his transmittal letter to the Secretary of the Interior of the 1967 annual report of the U.S. Columbia River Power System stating:

Perhaps the most disturbing disclosure resulting from load resource studies is the finding that on the basis of presently assured resources BPA will be unable to meet fully the power requirements projected for the Northwest's electro process industries. It is estimated that the Northwest will have to turn away over the next 20 years, seven out of ten potential new electro process industries due to an insufficiency of firm, low-cost power.

The current administrator, Mr. H. R. Richmond, in his letter of December 31, 1969, to Secretary of the Interior Walter J. Hickel, is even more pessimistic. Under the heading "Power Outlook," he says:

We are approaching a serious power supply situation in the Pacific Northwest over the next five years. Repeated delays in generator unit installation schedules at key Federal hydro projects in the region will result in utility loads outstripping resources during the years 1970–75. Table 2.2 shows the resulting deficits.

The deficit of 1,154,000 kilowatts shown for 1973–74 is equivalent to more than the combined peak loads of the cities of Tacoma and Eugene. Even with the elimination of interruptible industrial loads, the 1973–74 firm peak deficit will reach 262,060 kilowatts. This means that under certain conditions, we will be short of necessary reserves and it will be necessary to cut off firm loads.

[3] *Annual Report,* Bonneville Power Administration, 1969.
[4] *Statistical Year Book,* Table 33S, Edison Electric Institute, 1968.

We have not yet decided which firm loads will have to be dropped under these circumstances.

To consider another example, let us cross the Atlantic to England. What about the electric power situation in England? In Lecture 1, I stated that the net

Table 2.2 Peak Loads and Resources, West Group Area (thousands of kilowatts)

	1970–71	1971–72	1972–73	1973–74	1974–75
Load	20,159	21,472	22,777	24,076	25,449
Resources	19,734	20,992	21,757	22,922	24,840
Deficit	(425)	(480)	(1,020)	(1,154)	(609)

Table 2.3 Progress in Electric Power Supply in England and Wales

	1968/69	1947/48	Ratio 1968/48
Capacity Generation Stations (thousands of megawatts)	44,673	10,362	4.32
Generating Output (kwh $\times 10^6$)	173,418	36,391	4.77
System Generating Efficiency (Btu per kwh)	12,100	16,380	0.74
System Load Factor (percent)	52.7	48.4	1.09
Average Consumption per Domestic Consumer (kwh per year)	3713	1190	3.13
Average Price of Electricity Sold to All Consumers (pence per kwh)	1.914	1.144	1.68
Average Price of Electricity Sold to Domestic Consumers (pence per kwh)	2.14	1.297	1.65
Kwh per capita (1967)	3796		

Source: The Electricity Council Reference Section, *Reference Paper, Serial R.P.I.* (London: The Electricity Council, December 1969).

effect of the nationalization was that England has a power system that is unified and has had growth and performance that, on the whole, has given a good (but not particularly superior) account of itself. This can be seen from Table 2.3 that gives the highlights of the British power system over a two-decade period and Table 2.4, that gives the corresponding figures for the American utility industry—both public and private, but 77% private. Each of the tables gives in the fourth column the ratio of 1968 to 1948 performance. Whereas the capacity of the English stations grew $4\frac{1}{3}$-fold, that of

Table 2.4 Progress in Electric Power Supply in the United States Total Electric Utility Industry

	1968	1948	Ratio 1968/48
Capacity Generation Stations (thousands of megawatts)	288,750	57,300	5.04
Generating Output (kwh $\times 10^6$)	1,326,930	282,698	4.70
System Generating Efficiency (Btu per kwh)	10,300	13,650	0.752
System Load Factor (percent)	63.5	60.8	1.045
Average Consumption per Domestic Consumer (kwh per year)	6057	1563	3.87
Average Price of Electricity Sold to All Consumers (cents per kwh)	1.55	1.79	0.865
Average Price of Electricity Sold to Domestic Consumers (cents per kwh)	2.12	3.01	0.71
Kwh per capita (1967)	6616		

Source: EEI Statistical Year Books 1956 and 1968.

the United States stations grew 5-fold. The actual increase in generation was practically the same, but here it needs to be remembered that the British had a great deal to catch up on, on account of the long war interval through which her system had gone. Generation efficiency, expressed by system Btu per kilowatt hour, increased 26% for the British systems against 24.8% for the American systems, but the American systems ended up at a thermal performance of 10,300 Btu per kwh against the British 12,100 Btu per kwh—an almost 15% greater efficiency than the British. The average residential consumption increased 213% on the English systems but 287% on the American systems. The average price of all electricity sold in 1968 in England increased 68%, whereas in the United States it declined 13.5%. In the domestic field the rise was 65% in England, but there was a decline of 29% in the United States. And finally, in the kilowatt hours per capita in the year 1967 the British power system gave England and Wales an average kilowatt hour figure of 3796 whereas the American figure was 6616, that is, 74% above the British figure. It is quite obvious that the preliminary judgment in my first lecture is fully borne out by this statistical evaluation.

But if comparisons are made with a single private American system, the British figures do not look quite as good as in the comparison I have just made. I have chosen for the private system the American Electric Power System (AEP), representative of close to the very best in American privately owned power practice. Table 2.5 shows the comparison. Generating efficiency on the

Table 2.5 Progress in Electric Power Supply, American Electric Power System

	1968	1948	Ratio 1968/48
Capacity Generation Stations (thousands of megawatts)	9345	1966	4.75
Generating Output (kwh $\times 10^6$)	52,438	11,475	4.57
System Generating Efficiency (Btu per kwh)	9560	14,139	0.675
System Load Factor (percent)	72.8	67.8	1.075
Average Consumption per Domestic Consumer (kwh per year)	6652	1617	4.12
Average Price of Electricity Sold to All Consumers (cents per kwh)	1.03	1.39	0.74
Average Price of Electricity Sold to Domestic Consumers (cents per kwh)	1.87	3.01	0.62
Kwh per capita (1967)	9560		

Source: Annual Reports of American Electric Power System and author's files.

AEP System improved 32.5% against the British 26%, AEP load factor improved 7.5% against the British 9% but ended in 1968 at 72.8% against the British 52.7%, the ratio between the AEP and British load factors in 1968 being 1.38 to 1. The average AEP consumption per domestic consumer increased 312% versus the British 213%, the average price per kilowatt hour for all energy sold declined 26% versus a British rise of 68%, and the AEP residential realization declined 38% as against an English increase of 65%. And finally, the kilowatt hours per capita in AEP's service area were 9560, a figure 152% higher than the British 3796.

But if this evaluation appears slightly harsh, even though not intended to be, the British evaluation of the performance of their power system is even more harsh and, perhaps, unfair. The prestigious London *Economist,* in two important articles on the British power system, had this to say in two succeeding issues of that journal:[5]

> The power stations of England and Wales have a capacity of 45,000 megawatts, but on last Friday, as the snow came down, a "slight voltage reduction" of 3 per cent was ordered by the Central Electricity Generating Board because only 34,500 MW of actual power—77 per cent of the total—was available. It is true that November 28th was the trough of a premature cold spell which hit the southern half of Britain some three months earlier than usual—the third week of February was the last cold weather low. But this load shedding was a reminder of just how narrow a power margin is available to the 48.6 million people who depend on the CEGB for their electricity. . . .
>
> Because of what its engineers describe as basic and catastrophic faults in its latest 500 megawatt turbo-generators, the CEGB may have to move from the voltage reductions first described in *The Economist* last week to straight power cuts, with consumers being switched off at source, if the temperature drops to a point where demand swamps the narrow margin of peak-load generating capacity. The reason is simple: of eight turbo-generators each capable, in theory, of delivering 500 MW of power, only one is at present operating. . . .
>
> The CEGB got itself into this mess because it has been scaling up the size of its sets at a prodigious rate in order to get the economies of scale the big sets bring. New batches of designs were being ordered before an earlier generation had been installed, let alone run in.

[5] *The Economist* (London), December 6, 1969, p. 94, and December 13, 1969, p. 67.

The chart (not shown here) shows how the orders leap-frogged. . . . But in 1950 came the decision to order 100 MW and then 120 MW sets. This scaling up did not, however, represent any particular technical advance.

That came in 1953–54 when the CEGB decided to buy nine generators each capable of 200 MW. Here two technical leaps were made—which may in the long run prove at the bottom of this winter's grief. One was direct cooling of the copper conductors which act as the link between the revolving stator and the power supply chain, and the other, the increase of boiler operating pressures from 1,500 lb a square inch to 2,300 lb. From there on each increase in the size of generator was decided before the preceding batch of generators had been tried in service. Even now while the CEGB is wrestling night and day to sort out its 500 MW sets, the first of nine new 660 MW units are being installed. Yet, these units of, for Britain, unprecedented scale were decided on in 1965, two years before the first of the 500 MW sets were commissioned.

In making my observation with regard to the possible unfairness of *The Economist*'s harsh judgment, I need to call attention to two points: (1) any other power system management group, one in private enterprise for example, might be found that would duplicate a similar program of expansion, (2) such action could not with propriety be so harshly judged if the evil consequences of the action did not carry the potential of threatening the electric service of an entire country of close to 50 million people.

As for France, most of the French performance germane to this discussion was given in my initial description in Lecture 1 of the coming of nationalization to France. I do, however, want to offer a few figures on

Table 2.6 Some Performance Figures of Electric Power System of Electricité de France

System delivery in 1969	130.7×10^9 kwh
Assuming losses of 7%, sales in 1969 amounted to	121.6×10^9 kwh
Receipts in 1969 were	$13{,}352 \times 10^6$ francs
This is equivalent to	$\$2.4 \times 10^9$
Average realization was thus	1.98¢ per kwh
This compares with a realization by total U.S. utility industry in 1968 of	1.55¢ per kwh
And by American Electric Power System of	1.03¢ per kwh
In 1967 the French kwh per capita was	2287
And the U.S.A. kwh per capita was	6616
In 1969 the system input of Electricité de France was	130.7×10^9 kwh
For the AEP System the input was	62.7×10^9 kwh
Ratio	2.08:1
In 1969 the total personnel of Electricité de France was	94,810
For the AEP System the total personnel was	14,153
Ratio	6.7:1

Source: From a private communication from one of Electricité de France's technologic administrators.

French performance in 1969. These are shown in Table 2.6.

Table 2.6 is an interesting table. Perhaps the most interesting item, and I think a very important commentary on nationalization, are the figures on personnel. With a ratio in energy system input between the French and a specific American power system in 1969 of 2.1 to 1, the ratio of personnel was 6.7 to 1.

Let us consider Sweden. I have already indicated

my judgment in the first lecture about the remarkable integration of public and private power that has been carried out in Sweden. The chairman of the Swedish State Power Board, in discussing the latest figures of costs of energy for industrial purposes in the fiscal year 1970, gives the tabulation of Table 2.7.

It is very interesting also to consider the comments of the chairman of the Swedish State Power Board on nationalization. He says:

> The question of nationalization of the power industry has from time to time come up to discussion. The premises would appear to be advantageous as there is a state utility responsible for nearly half the total production and which in addition owns the whole trunk line system for higher voltages.
>
> That nationalization has not been done and is not regarded as very probable in the future can be attributed to the following general evaluation. The benefits which, as regards planning and technical questions, might be won through nationalization would be more than offset

Table 2.7

	Average Current Price for Industrial Electricity (mills per kilowatt hour)
Norway	3.29
Sweden	7.74
United States	9.08
France	13.54
United Kingdom	15.48
West Germany	16.8
American Electric Power System for 1969 [added by the author]	7.44

by the disadvantages resulting from the elimination of competition.

One of the premises for a mixed system with a competitive element is of course that the different utilities compete under equal conditions. [If great care is taken in the setting up of the conditions. P.S.] The State Power Board has accepted this fact. Although the Board receives the capital by parliamentary appropriations and not from the open capital market, the Board is required to show a profit which would correspond to that normally expected of a private utility with 25% of its investment capital contributed by common stock. In the profit expected shall also be included a compensation for the taxes a private utility would have to pay. The result is that the national utility can compete with the others for customers without giving rise to political or ideological conflicts.

The competition situation has necessitated rationalization measures and has during the 1960's resulted in price cuts at a time when the majority of other products has increased in price.

This to me represents a remarkable example of combining democracy, economics, and social ethics.

I shall not pursue these statistical evaluations of technical competence and performance for Mexico or Italy or Russia, even though I believe they would prove most interesting. But I shall have occasion to make some observations in this area in my next lecture. I do, however, want to make a few observations on research and development as influenced by social organization.

R

RESEARCH AND DEVELOPMENT

Essentially, in the United States, the utilities both

private and public have never developed a tradition in or a well-rooted policy for research. Instead of a policy we have rather settled down to a practice expressed by "Let George do it," and "George" has been one or two maverick private utilities and, in their sphere of responsibility, one or two of the manufacturers. Some research work is being carried out centrally by a group styling itself as the "National Research Council." But it is supported on any project it tackles only by those utilities expressing an interest in that specific project and its influence to date has been minimal.

In the case of atomic power, we have been fortunate to have had the strong interest and influence of both the Atomic Energy Commission and the Congressional Joint Committee on Atomic Energy, both of which have stimulated and supported a great deal of research, particularly in reactor prototypes, fuel, waste disposal, and safety.

In a few areas, publicly owned utilities in the United States and abroad have done some effective work: Bonneville in extra-high-voltage direct current; TVA in hydraulic design and construction; Electricité de France in extra-high-voltage alternating current; the Swedish State Power Board in extra-high-voltage alternating current and in direct current; and the British Central Electricity Generating Board in gas-cooled reactors. But it all adds up to a pitifully small amount when you consider the magnitude and the scope of the problems facing the electric supply companies throughout the world.

The failure of the privately owned companies to do more in research is a great pity. Basically they are in far better shape to carry out effective research than the public power companies. The greater continuity of their managements, the much greater flexibility in setting up ad hoc organizations, the greater freedom they possess in bringing in manufacturing organizations as partners, and finally their ability to carry through most of the appealing projects on a regional basis give the private sector almost unmatchable advantages to simplify the organization and implementation of significant research.

As commentary on the foregoing, I would like to cite the experience of the AEP System. In a period of 45 years, that system, working with manufacturers on some ideas and without them on others, brought into being solid advances in generation of coal-fired steam-electric power by the employment of reheat, by completely revolutionizing the cooling of electric machines by substituting hydrogen for air, by increasing size of boilers and turbines from 10,000 kw to 800,000 kw and power plants from 30,000 kw to close to 3 million kw, by increasing temperatures of steam from 350° to 1050°F, and by increasing thermal efficiency of units and plants from approximately 15% to 40%.

In the area of environmental accommodation and development of compatibility between electric generation and minimum burden on the environment, the AEP System pioneered and helped bring into being two important developments in large-scale generation: the natural air-flow large-size wet cooling tower[6] and the tall stack.[7] As far back as 1952 this organization adopted and publicized the tall stack as the one available solution to the problem of diffusing SO_2 concentrations to the point where they would become entirely harmless to life at ground level, whether vegetable, animal, or human.

These practices not only had a very telling influence on the growth and development of the AEP System but influenced the art and technology of power generation in the United States and, indeed, the world over.

In transmission this same organization carried through a continuing program of research and development of higher and higher voltages, raising national levels from 287,000 to 345,000 and 500,000 volts and subsequently—an international high—to 765,000 volts, and in doing so improved the reliability, economics of, and the infringement on the landscape by transmission of electric energy.

Many ingenious and sometimes highly inventive developments contributed to make these transmission improvements possible. Among them was a principle enunciated, and subsequently universally adopted, of insulation coordination, the essential element of which is the idea that in a power system the orders of strength

[6] Howard C. Barnes, T. T. Frankenberg, and C. P. Lugrin, "Highlights of AEP Big Sandy Plant," *Power Plant Engineering,* December 1962.
[7] D. H. Williams and J. T. Dowd, "Design and Construction Features of the 1600-MW Mitchell Plant," American Power Conference, April 1969.

of the various links in a chain had to be in proper relationship one to another, and, while for different purposes different insulation levels might be required to give adequate service, all the links in the chain would always be properly coordinated at any level chosen.

The practice of ultrarapid reclosure of circuit breakers was researched and developed on the AEP System. This was based on the philosophy that the function of a circuit breaker was not merely to interrupt a circuit when it was in electrical difficulty but to interrupt it and then to reclose it. Both operations are carried out so quickly that not only was the circuit fault cleared with minimum damage to faulty equipment but service was restored so quickly that the effect was as if the interruption had never happened. This, too, is now the worldwide standard practice.

This organization brought into being one of the most efficient, if not the most efficient, integrated power systems in the world to supply the electric power needs of its service area and provides a grade of service equal or superior to the best metropolitan standards to each of its 2256 communities, all of relatively modest size.

As a result of complete integration and coordination of facilities, aided by the almost completely exhaustive exploitation of interconnection now carried out at about 90 points on the periphery of the system at voltages varying from 138 kv to 500 kv, it has been able to provide service at an average kilowatt-hour price lower than that of any other private- or investor-owned system. Thus it has set high standards for others to follow and to improve upon.

In the domestic field the organization began research as far back as 1931 on the development of the heat pump, which not only is a great contribution to comfort and complete electrification of the home but also has in it the seed of a reduction of total heat rejected to the environment, and thus can help solve a thermal environmental problem that could be difficult with the further growth in the globe's population.

All this work, and a great deal more that one hesitates to go into detail in this lecture, was made possible largely by dedicated adherence to the principle of improvement by seeking new and better ways, searching and researching in other words, and doing it with a determination and persistence. This effort did not have to concern itself with changes in political administration or indeed changes in corporate administration, because these were practically nil over an extensive enough period to make continuity of ideas and developments bear the maximum fruit.

This lengthy citation is given not with a view to establishing a thesis that a system of competing private utilities in power is going to result in outstanding management, as exemplified by the achievements of the AEP research program, in all such private power organizations, but merely to suggest that it is more likely to produce a few such AEPs than is a system of public ownership with its rigidities and tendency toward monolithic control.

Thus, it would appear that a careful weighing of experience of the pros and cons of privately and publicly owned power supply, from the standpoint of developing higher technical competence and performance, would in total show a plus balance in favor of private organization. This is because of the continuity it offers in organization in pursuit of important and long-term ends by providing freedom from interference (except in the case of incompetency) and because of incentive.

Even this advantage does not tip the scales heavily and irreversibly in favor of private enterprise. Private enterprise in electric power can and does develop basic weaknesses, particularly after a period of prosperous operation, when there is a tendency for top organization to neglect to develop continuously prudent and improvement-seeking habits. It is easy to make a profitable showing, or so it frequently appears, without such Spartan discipline. In fact, directors, who should be the continual watchers on the towers of performance and corporate health, can cease to direct management and become directed by management.[8] As *The New York Times*[9] recently pointed out in connection with an analysis of the tragedy of the Penn Central collapse, which had been covered in the August 1970 issue of *Fortune Magazine:* "scores of men who were charged with the fiduciary responsibility of supervising the management of Penn Central had failed miserably." The same article states that, "Fortune's accusations create some interesting legal questions, primarily those of who should be sued by the shareholders, management or the board."

[8] I have dealt with this subject at greater length in "Technology, Environment, Morality and Ethics in the World of Today and Tomorrow," an Honors College Lecture delivered at Wesleyan University, Middletown, Conn., April 15, 1970.

[9] July 26, 1970.

The shareholders have nothing to do with the selection of the management. They elect the board of directors and by this act entrust their property to that body. The board has the responsibility of finding a competent management and then deputizing it to carry on the affairs of the corporation on a day-to-day basis while the board continues to set policy and give guidance and direction to the management. But the U.S. system at the present time has no way to assure itself that directors truly direct. This inherent weakness in our business society is in bad need of curing in order to make private enterprise a fully responsible operation and to assure its success.

Public organization of power, too, has its weaknesses: By avoiding its citizenship duties, and helping its consumers avoid theirs, of supporting the economy through taxation in such areas as defense, education, health, and similar important socioeconomic operations, it undermines these activities. Because it has relative freedom from control of any kind (on the theory that, being publicly owned, it cannot possibly do anything wrong, which is both a fallacy and contrary to human experience), it tends to make its policies less than completely responsible. Because there is no higher authority, public power carries with it the seed and germ of corruption: It leaves too many people in the public power employ with a smug feeling of "where else can they go if they do not like what we do?" This is corruptive, and will be as long as human society is corruptible.

This examination of the technical competence and performance of the public and private forms of power still does not give us the answer to the question: Which is the best form of organization for a country's electric power supply? One of the questions to probe is: Which kind of organization can best protect the two important interests, those of the consumer and those of the public? This aspect will be examined in Lecture 3.

3

Lecture Three

PROTECTING THE CONSUMER INTEREST AND THE PUBLIC INTEREST

The consumer interest in the performance of the electric power supply industry is very important. It is not the only interest, even though the consumer in general pays all or almost all the costs of the supply—at least in a private supply. Because when the power supply is utilized, it has broad-range effects on society, on the rate of its growth, on its economic welfare, on its social welfare, and, on occasions when the society is challenged, on its safety and its ability to survive attack. Thus, the public interest is also very important.

A

AREAS OF CONSUMER INTEREST

There are four principal areas of performance in which

the consumer interest is demonstrated: availability, reliability, price, and a continuous program of planning for the future. Let us examine each of these areas.

A

Availability

Over the 88-year history of the electric utility industry, a definite pattern of performance with regard to availability has been established among the leading societies of the world, and this is shown clearly in Table 3.1.

In my first lecture, in discussing the Mexican situation I stated that while nationalization improved the Mexican power situation enormously, it still did not give the Mexican people a satisfactory power supply system, even though the government has actively intervened in this area for the last 20 years. In order to discuss the criteria for this judgment, I have prepared Table 3.2 as a supplement to Table 3.1. In it I have shown ten smaller electric energy producing countries of the world which in 1967 ranked lower in total electric energy production, and I have listed them in the order of their total energy production. I have also shown the kilowatt hours per capita, and the rank in kilowatt hours per capita among the countries listed. Of these ten, Mexico's position is tenth—last—with a most inadequate kilowatt hours per capita of less than 500. It has a long road to travel before reaching a destination that can be designated as an electrified country.

Table 3.1 gives the record of 12 countries as of 1966 with regard to population and electric energy produc-

Table 3.1 Population, Electric Energy Production and Utilization, 1966

	Population (millions)	World (%)	Electric Energy (billions kwh)	World (%)	Per Capita (kwh)
World	3355	100.0	3513	100.0	1050
China (Mainland)	709.5	21.3	40	1.1	56.5
India	497.7	14.8	38.4	1.09	77.0
Soviet Union	233.1	6.8	516.5	16.1	2215
United Kingdom	54.62	1.62	190.0	5.41	3470
United States	196.3	5.7	1249	35.5	6360
Indonesia	106.4	3.1	—	—	—
Pakistan	104.9	3.05	—	—	—
Japan	98.86	2.87	210.1	6.0	2130
Germany (West)	54.91	1.63	166.3	4.75	2865
Canada	19.98	0.60	157.4	4.48	7900
France	49.26	1.47	106.1	3.02	2170
Italy	51.91	1.54	87.1	2.48	1685

Source: *The New York Times Encyclopedic Almanac*, 1970.

tion. It is interesting that the United States, with 5.7% of the world's population, produced 35.5% of the world's electric energy. The nearest to it in this performance was the Soviet Union which with 6.8% of the population produced just slightly over 16% of the electric energy. As against this, the mainland of China, constituting over 21% of the population, produced just slightly over 1% of the electric energy. India, with close to 15% of the population, just about equaled China's electric energy production.

There cannot be any question that the American

Table 3.2 Electric Energy Production and Utilization of Ten Smaller Electric Energy Producing Countries, 1967

	Electric Energy (billions kwh)	Rank in Production	Per Capita (kwh)	Rank in Per Capita (kwh)
Belgium and Luxembourg	27.22	1	2745	3
Switzerland	24.44	2	4026	1
Romania	22.69	3	1176	7
Mexico	21.06	4	461	10
Austria	19.76	5	2698	4
Yugoslavia	18.84	6	944	8
Finland	16.75	7	3571	2
Argentina	16.63	8	715	9
Hungary	14.38	9	1407	6
Bulgaria	13.67	10	1656	5

Source: Felix Fremont: "Annual Growth Rate on Downward Trend," *Electrical World*, July 6, 1970.

power supply system, 77% of which is private, has in comparison more than held up its end and indeed has turned in an outstanding performance—superior to that of Russia, the United Kingdom, France, and Italy, in each four of which electric power is completely nationalized.

R

Reliability

During World War II, there was great concern with

brownouts and blackouts, but the only substantial brownouts were due to military requirements for security. Otherwise, all civilian and military requirements were met throughout the four years of an unprecedented war.

There was a good deal of talk about brownouts during the postwar period, but these did not materialize. The only significant outage was the Northeast blackout of 1965—much more severe than a brownout. Without question, this was bad performance. It was not a failure due to inadequate technology. Essentially, it was due to bad system planning. The fact that this weakness may still not be entirely cured is essentially a fault of utility management and regulatory body action or bad action, mostly at the Federal Power Commission (FPC) level. Here the weakness is compounded of a mixture of ignorance and lack of initiative.

Given the willingness to do so and the analytical efforts this involves, every power system can be tested in advance of any major change or development in its composition, for example, whether it will or will not be stable in case of a major emergency. There is no shortcut to that basic fact nor to the requirement that changes necessary to avoid an unstable condition must be made mandatory. However, the will to carry out these very difficult changes is invariably attenuated by the irresolution springing from political considerations.

Some of the same considerations are responsible for the current very bad situation in the city of New York, even though there is no question about the integrity of the management or its deep desire to do everything that is necessary to cure the difficulty. The New York company and its top management are responsible for seeing that the New York system maintains a balance between the power supply and the demand imposed on it, and they have been driven almost frantic by repeated irresponsible disregard of various community segments for anything other than their own limited interests: The aesthetic environmentalists are willing to have the supply taken care of by steam-electric energy fueled by fossil fuels or by atomic fuels but not by hydroelectric means. The atomic environmental viewers-with-alarm are perfectly willing to have the power supply taken care of by hydroelectric or steam-electric means, fueled by coal, oil, or gas but not by nuclear fuel. The apostles of air unpolluted are perfectly willing to see energy produced by hydroelectric or nuclear electric means but definitely not by fossil fuels.

It is very interesting, however, to find that even in completely nationalized power systems these problems have not disappeared. In England, for example, all energy supplies are publicly owned, with the exception of oil and some natural gas. At the moment they are all the responsibility of the ministry of technology. But the problem of developing the proper mix of primary energies in the electric power supply is in great disorder. According to *The Economist* (London) of July 27, 1970, pp. 58–59:

> Oil is taxed in order to equalise what would otherwise be an overwhelming price advantage over coal, gas,

electricity and nuclear power, and now coal, gas and electricity are all seeking price rises, which will, of course, increase the competitive advantage of oil. But now that Britain is developing its own local oil and gas supplies in the North Sea, should the other fuel industries continue to get the same degree of protection?

If they do not get the price rises being asked for, coal, for one, will take a bigger loss this current year than the £50 million allowed by Parliament. The coal board's difficulty is that it cannot step up productivity faster than 3 per cent a year, while costs are racing ahead probably three times as fast. The coal board says it is a matter of getting coal-cutting machinery to work for rather more than an average of two hours out of every seven-hour shift. The rest of the time goes in breakdowns, power failures and accidents. The board knows every extra minute that the machines can be kept working is equivalent to another £4 million of coal dug. But the Central Electricity Generating Board could make exactly the same sort of excuses for the breakdown last winter of its new generating plant. The question both these industries need to be asked is what went wrong with their buying policy that they placed so much confidence in apparently untried designs bought off the drawing board. There have also been serious errors in forecasting electricity demand, which has in turn affected the scale of the CEGB's investment programme. Of course, had the board not over-estimated demand, the plant breakdowns would have caused bigger disruption to supplies last winter than in fact they did.

Not such a good management record, but still better than what has been going on in nuclear power, where the commanding reputation built up by Britain in the 1950's has been allowed to melt away, with the design resources accumulated in the Atomic Energy Authority scattered to the winds among industrial groups that simply have not the capital resources to take up the work where the authority left it off.

P

Price

Heated figures, statistics, and facts have been presented on the advantages of public power based on the supposedly indisputable fact that public power means cheaper and, therefore, lower-cost power. Figures are cited, for example, on the cost of 200 kwh of residential electric energy in the city of Los Angeles as against the city of New York, or the cost of industrial power for a block of 2000 kw demand for a chlorine plant at Calvert City, Kentucky, versus Saltville, Virginia.

At such times, as one attempts to unravel the intricacies and booby traps involved in making such comparisons of costs, one is almost moved to accept as economic wisdom Sydney Smith's dictum on facts and figures, "don't tell me of facts, I never believe facts. Canning said nothing was so fallacious as facts, except figures." Yet, while mindful of the booby traps, I would like to present Table 3.3, of facts and figures.

Table 3.3 shows consumer prices, wholesale commodity indices, average residential price of electric energy, and *total ultimate consumer price of electric energy;* all these are for investor-owned utility companies extending over a period of 14 years. While consumer prices went up 30% and the wholesale commodity price index 17%, the average residential electric costs went down 18.5% and the *total ultimate consumer price of electric energy* went down over 11%.

When looking at these statistics, one cannot help

Table 3.3 Investor-Owned Utility Consumer Price versus Commodity Price

Year	Consumer Prices 1957–1959 = 100	Wholesale Commodity Index	Average Residential Price (¢ per kwh)	Total Ultimate Consumer Price (¢ per kwh)
1955	93.3	93.2	2.76	1.80
1960	103.1	100.7	2.62	1.82 (high 1.85)
1962	105.4	100.6	2.56	1.80
1963	106.7	100.3	2.52	1.77
1964	108.1	100.5	2.45	1.73
1965	108.9	102.5	2.40	1.70
1966	113.1	105.9	2.34	1.67
1967	116.3	106.1	2.31	1.66
1968	121.2	108.7	2.25	1.64
Change %	30.0	16.6	−18.5	−11.3

Source: Edison Electric Institute, *Statistical Year Book of the Electric Utility Industry for 1968.*

asking whether these electric energy figures might have been lower under public organization. The obvious answer is that under public organization these electric energy figures *might* have been lower, but it does not follow that costs *would* have been lower. Everything else being the same, it might have actually been better for everyone—the consumers, the public in general, the investors in utility companies—if costs had been raised and prices had been raised above the level shown in Table 3.3 by an excise tax of, let us say, 0.15 mill per kwh that would have yielded an annual fund of some $200 million dedicated to research.

I want to explore the very important questions of costs, subsidies, and prices. Some three years ago my colleague, Abraham Gerber, and I presented a paper[1] that could be incorporated in this lecture almost completely. Although I shall pass up this temptation, I shall quote at some length from it, since it is so appropriate to the matter at hand.

> The major source of error in the economic evaluation of water resources projects most frequently stems from the use of an incorrect, usually too low, cost of capital. Economic analyses frequently are made on the basis of costs of capital associated with the interest rates on government loans within a country, or intergovernmental loans, or the interest rates on loans from specialized international development agencies rather than on the real cost of capital, or in technical terms, the opportunity cost of capital in the particular economy. In this it is necessary to draw a distinction between real cost and money cost. . . .
>
> From the point of view of optimum development and use of resources, the comparison of the costs of several alternatives needs to be based on true representations of the various alternative costs in terms of real resources rather than on distortions based on artificially contrived monetary arrangements. This can perhaps be illustrated best by a real example.
>
> The United States Federal Power Commission periodically prepares appraisal reports for licensing various potential hydroelectric projects. In one such report, the Commission, using the exact same capital cost for the

[1] "The Social-Economic Evaluation of Water Resource Projects," presented at International Conference on Water for Peace, Washington, D.C., 23–31 May 1967.

exact same project, arrives at the following annual costs:

By Private Financing	$4,442,000
Financing by the Federal Government	1,507,000
Financing by the Rural Electrification Administration	1,370,000

We note here three annual costs in descending order from private financing to Federal financing to Rural Electrification Administration financing. The Rural Electrification Administration obtains its funds from the Federal Government at a fixed interest rate of 2% irrespective of the cost to the Federal Government of borrowing such funds. In this particular case, to determine the cost to the Federal Government an interest rate of 3.125% was used. For private financing an interest rate of 6.25% was used. In the case of the Federal Government, no additional cost resulting from taxes forgone was included. In the case of the Rural Electrification Administration, .5% was allowed for taxes or payments in lieu of taxes. For private financing 5.72% was included for Federal and all other taxes. In all cases the annual benefits, based on the cost of an alternative privately-financed thermal power plant, were exactly the same—$2,648,000. These benefits were based on the alternative cost of a privately-financed steam-electric power plant. Thus the benefit cost ratios were: for REA financing 1.93; for Federal financing 1.76; and in the case of private financing 0.6, or less than 1.00.

This, therefore, allegedly demonstrated that on the basis of private financing the project was economically infeasible. Thus, with exactly the same installation, the same use of real resources, and the same benefits, the economic desirability of the project varies not only in the degree of desirability or viability, but even to the extent of changing it to undesirable or nonviable, depending on what choice of financing one selects. Furthermore, one is confronted with the absurd "logical" conclusion that an agency which, in the economic sense, is a ward of the Federal Government, borrowing money from the Federal Government at what constitutes in effect negative interest, can do the job at a lower cost than the Federal Government itself. This obviously irrational conclusion, stemming from an Alice-In-Wonderland system of economics, did not in the least daunt the Commission. One can only conclude that this is another illustration of the failure to fully comprehend the real cost of a project to the economy. . . .

In the case of this particular Federal Power Commission analysis, if in fact power can be sold up to the full capability of the project at a price equal to the cost under private financing, and assuming that private financing reflects the market forces reasonably well, then the project costs must be measured on the basis of the return including taxes on the privately-financed project. The free market rate of return, to the extent that this is determinable, is the proper element to employ in evaluating costs rather than some arbitrarily determined governmental rate.

TAX ELEMENT NEEDS ANALYSIS

The tax element needs further probing and the analysis also. It will be noted that in this Federal Power Commission illustration the tax component for the privately-financed project is almost equal to the rate of return—5.72% compared with 6.25%. It has been suggested that if such taxes were excluded from the computation of the cost of all alternatives, the right answer would be arrived at. The particular taxes in this case are Federal Income Taxes. In the case of import duties and similar ad valorem taxes it is appropriate to exclude such taxes from the computation because they do not represent real costs to the community and their discriminatory effects can distort resource allocation. In the case of income taxes, however, these do represent real opportunity costs, since they represent part of the return that is considered throughout the economy in making an investment decision and they thus represent part of the alternative returns available. The fact that part of the return is distributed to the investor and part to the tax collector merely represents the allocation of the return on the investment between the investor and the government. Therefore, the return, including income taxes, represents the true investment opportunity cost to the economy. To put it another way, the tax component is that part of the productive return on the investment that is allocated to support government functions.

A Soviet economist, Z. F. Chukhanov, Corresponding Member of the USSR Academy of Sciences, arrived at this same conclusion, although he expressed it slightly differently in a paper that he prepared a few years ago for *Teploenergetika.* Of course, it was necessary for him to contrive an artificial substitute for a market rate of interest. However, he also allowed a rate equivalent to taxes, which he described as that income from investment necessary to meet the nation's "quite large nonproduction expenses for pensions, scientific research, defense, etc. . . . Since power generation, as most other branches of the national economy, must contribute its share to the national income, it is necessary to add another term to (the social cost) equation." Mr. Chukhanov, in his paper, by taking into account capital cost including the equivalent of taxes for support of the government, concludes that hydroelectric development in the Soviet Union has been wasteful when compared with the alternative of thermal power stations. Further, he concludes that in the period 1952 to 1958 no hydroelectric power capacity should have been built in the Soviet Union. The strong shift in emphasis in Soviet power development in recent years from hydroelectric to thermal indicates that this basic argument has had some effect on Soviet thinking regarding water resource development.

This points up another element. A soundly carried out economic analysis is independent of the particular economic system. It applies to a system in which all facilities are owned by the state, to those in which no facilities are owned by the state, and to all shades and variations in between. Thus, in the case of Great Britain, with a mixed economy, but one in which the national government owns all the electric power resources, a "white paper" published in 1961 pointed out, "If the profitability of capital development is assessed on different (and easier) financial criteria from those adopted in industry generally, there is a risk that too much of the nation's savings will be diverted into the nationalized industries. Again, if the prices of the goods and services which the nationalized industries provide are uneconomically low, demand for them (and for investment to produce more of them) may be artificially stimulated. Thus the operation of the nationalized industries with an unduly low rate of return on capital is sooner or later damaging to the economy as a whole." This "white paper" further pointed out that British industry as a whole is earning a gross rate of return, including taxes, of about 15%. However, it stated that because utility service involves somewhat less risk than other industry, the nationalized utilities can earn a somewhat lower rate, and a figure of about 12½% for the return on Central Electricity Generating Board facilities was established.

In my closing discussion of our paper, I said:

There are any number of schools of economics. Mr. Norwood is, obviously, a member of the social-religious school. In his paper, to use his own words, he "commends the United States Government policy of using low-cost money to achieve optimum development of water resources."

One comment I can make on this view is the judgment given to me some 17 years ago by a principal manager of the nationally owned water supply system in Israel, that the capital facilities of that system ought not to be burdened with any interest charges whatsoever in view of the great social importance of water to that society. Interestingly, and not surprisingly, the Minister of Finance of the country disagreed completely with that view.

The trouble with such a philosophy, admirable as the advanced social-religious views behind that may be, is that it is also based on the most primitive notions of money as something that any sovereign government can make available at high or low cost or no cost whatsoever or even at negative cost, without transfer of heavy burdens to other segments of that society, and that this can be done without limit. It disregards the hard fact that money utilized constitutes a demand upon or an allocation of a society's limited or finite resources: materials, research, fabrication plants and human labor—unskilled, semi-skilled and highly skilled.

The mechanism of the price of capital funds constitutes a grand market queuing operation for national resources and provides one of the most rational means yet devised for determining at a particular time what gets built, when it gets done, and at what cost.

No government can determine the interest rate it will pay for funds it borrows. The other day the Secretary of the Treasury of the most economically potent and productive nation in the world, Secretary Fowler, asked Congress for removal of the 4.25% ceiling on (long-term) government bond interest rates. [Please recall this was originally written some three and a half years ago. P.S.] He had found, what Congress had apparently chosen to disregard when it set the upper interest rate limit, that the Treasury has frequently not been able to find enough buyers at the fixed-interest figures to make bond offerings possible. This complicated enormously the job of the Treasury and in other respects resulted in adverse effects on the country's economy.

What Mr. Gerber and I have tried to demonstrate is that a sound economic evaluation of water resources projects is independent of the social, political or other motivations, and is independent of the particular economic system on which a society is organized. If a given society is not to be led astray and if it is not to make a mess of the indispensable business of properly allocating its limited total resources, it is important that the proper

—and this means total—costs be used in the evaluation. Having done this, the society is then in an excellent position to assign priorities. This does not prevent it from upgrading the priority of any socially or defense-desired project at the expense of another less costly. But the intelligence and sound judgment with which this will finally be done will always be materially enhanced by having properly determined values and costs. Subsidies, desirable and granted, do not change cost.

Thus, taxes forgiven in the cost of electric energy, are taxes lost. They are a subsidy. But electric energy in a modern society, thanks to its continuous and sound technological development over almost 9 decades, is such a great economic buy that it needs no subsidy. If the figure for federal, state, and local taxes paid by investor-owned utilities is taken at 3.8 mills per kwh,[2] this subsidy, if it were given impartially to all energy users, would in the year 2000, with an expected distribution of 7.2 trillion kwh, represent a sum of $27.5 billion per year. Surely that is a source of funds of a magnitude that no society will be able to wave aside as being of no moment.

Thus if the basic difference between public and private power in the cost of money and avoidance of taxes disappears, where else can there be a saving in cost of public power? In more efficient organization? In more efficient operation? One can ask these questions, but in the light of earlier discussions of organization and quality of people, one is bound to dismiss them as offering no basis for acceptance.

[2] *EEI Statistical Year Book 1968.*

P

Planning for the Future

Here again the greater freedom under private enterprise, the much greater continuity, and the much more substantial rewards in position, in money, and in status (particularly with the absence of the inevitable political denigration of public achievement), private enterprise offers much greater opportunity to plan far into the future with good prospects of implementing carefully made plans. Finally, under private enterprise there is bound to be a much greater diversity of resources put at the disposal of those who show a capability, particularly when competence and integrity are once established.

A

AREAS OF PUBLIC INTEREST

In the first lecture I alluded to testimony that unreliability of performance by members of the business community failed to take into account the public interest. Disregarding the smugness in this exclusive wrapping of the mantle of concern for the public interest around the shoulders of the members of a federal regulatory staff, one can but wonder at the imperceptiveness of the brand stamped on our citizens in general and the implied national tragedy if this were true. Fortunately it is not true, and I myself, not being ready to accept this defamation, propose at least to examine the views, pro and con, about the ability of public power to take care of the national interest.

In a way, I have already partially answered this. To the extent the consumer interest is taken care of at least as well by private power, the public interest is likewise equally well covered. The question then has to be answered in terms of a possible national crisis stemming from a national catastrophe such as a drought, or an upsurge in national defense activities, or war itself, or in a critical postwar period when the maximum co-operation and coordination of the various parts of a country's power facilities have to be achieved to serve the national interest properly. To answer this question, one can call upon the experience preceding, during, and following World War II.

Coordination of the operation of generation and transmission systems is almost as old as the industry itself.[3] First came the coordination of a single system and later a certain degree of coordination between systems to effect savings in generating reserves and other economies.

By the middle 1930s the "tools" of operation and the techniques of interconnected operation had developed to the point where widespread interconnection was possible. The tools consisted, in the main, of high-speed relaying, high-speed circuit breakers, ultrarapid reclosing of lines, improved excitation systems, improved lightning protection, and automatic frequency–tie-line

[3] I am indebted to my former colleague, V. M. Marquis, retired vice president and assistant to the president of American Electric Power Service Corporation, for most of the material in this discussion.

control. It was the development of frequency control equipment and later automatic frequency–tie-line control—that made possible widespread interconnected operation.

Fortunately, these advances had been made by 1939 when the defense period began. It made possible the operations throughout the nation that led to ample power for defense and war production, with the minimum use of strategic materials in short supply.

Examples of how electric utility systems, regardless of ownership, can coordinate with outstanding results, especially during national emergencies, can be listed as follows:

1. The Southeast drought emergency in 1940–1941; mobilization of power supply from surrounding areas.
2. World War II experience; the high degree of coordination during World War II among utilities in the same regional or geographic areas.
3. Postwar conversion period; coordination of operating and construction schedules after World War II during the conversion from war to peace and to meet the unexpected peacetime boom demands for electric service.
4. The Korean situation; coordination to meet defense and civilian requirements.
5. The supply of interim power to the AEC Oak Ridge Diffusion Plant in 1951; coordination of surrounding utilities in 1951 to furnish interim power to the expanding AEC gaseous diffusion operation at Oak Ridge.
6. The OVEC story; the two sets of utility groups

organized to furnish, first interim and finally permanent, power to the AEC diffusion plants built at Paducah and Portsmouth.

While it is true that all the cases here listed deal in the main with national emergency conditions where all affected organizations are anxious to help (knowing all the time that if they do not cooperate they will be ordered to do so), nevertheless the cases clearly demonstrate the high degree of cooperation that can be accomplished. Under more normal conditions, the problem becomes more difficult because of the multitude of problems faced in getting a large group of individuals to work together in the absence of the unifying national emergency.

I would like briefly to comment on the cases listed.

T

The Southeast Drought Emergency 1940–1941

In 1940 the Southeastern states had a generating capability of some 4050 mw, of which 66.5% was in hydroelectric plants. The average for the country at that time was about 28% hydro.

A severe drought developed in 1940 and extended through 1941. Rainfall was exceedingly low, and reservoirs were drawn to dangerously low points. At the same time, to meet defense needs the demand for aluminum and other similar products had skyrocketed.

The surrounding utilities began at once to furnish power to the stricken area. In May 1941, the FPC entered the picture on the basis that an emergency existed and ordered certain measures, including the building of several interconnections with surrounding companies. The FPC intervention did not work too well, and, in the fall of 1941, the Power Division of Office of Production Management (OPM) was asked to enter the situation. Their first step was to put knowledgeable experienced people in charge, and things improved at once. While some curtailment and a brownout were necessary for a short time, there was no slowing up of the defense effort in this area.

It had been estimated that some 32 million kwh weekly could be delivered into the area from the outside. However, the extremely close coordination resulted in actual weekly deliveries of 42 million kwh. (The actual quantities of energy delivered must be judged against a background of 30 years ago, when utility systems of the country were roughly one-eighth of their present size.)

A great deal was learned from this experience both by OPM and the utilities and was most helpful in the next great emergency, World War II.

W

World War II Experience

While the Southeast incident applied to only one corner of the nation, World War II meant the marshaling of power facilities all over the nation to meet the new, as well as existing, war loads with the minimum use of scarce materials.

Steps were immediately taken to coordinate power supply in each region under the guidance of the Power branch of OPM.[4] The Power Division recruited expert personnel and brought them to Washington to assist in this effort.

It soon became apparent that this problem of co-ordination could not be done from Washington, and the Office of War Utilities (OWU) very wisely followed the principle of urging interconnected utility groups to form coordinating committees to make sure that the interconnected systems were properly coordinated. One of the duties of the committees was to furnish accurate capacity and load data to the staff in the Office of War Utilities who were assigned the task of following closely the various power areas throughout the country.

Prior to the war, considerable work had been done on tightening the operations of certain interconnected groups throughout the country, and these were selected as proper coordinating groups during the war period.

The country was divided into eight great power areas. Actually, during the war four of these groups, namely, the North Central, the Illinois-Wisconsin, the Southeastern, and the Southwestern areas, all operated in parallel with an interconnected demand of over 16 million kw (now well in excess of 100 million kw). Like-wise, there was coordination between the New England and New York groups and the Pennsylvania and New Jersey groups. These coordinating committees func-tioned quite differently in various sections of the coun-

[4] OPM later became WPB (War Production Board), and the Office of War Utilities was set up under WPB.

try, but in general aimed for and accomplished the same results.

During this period the power systems of the country were called upon to do the almost impossible. From 1939 to 1944 war demands increased the requirements 73% from the generating plants in the nation while the generating capacity of all these plants during the same period was allowed to increase only 26% in kilowatt rating. Standby equipment and all possible emergency reserves were put to work. Plants and lines were overloaded beyond the point that it had been believed possible previously. Generating reserve standards set by the Office of War Utilities were 8% for steam-electric systems and 4% for predominantly hydro systems. All of the nation's power systems—private, municipal, county, state, and federal—were in effect assembled in great operating pools. Maximum use was made of all existing facilities, and new equipment was allowed to come into operation only where it would be of the greatest overall benefit. A large aircraft carrier required turbines rated at 450,000 hp, a battleship 225,000 hp, and lesser amounts on down the line. The turbines for these ships, so badly needed for the war effort, had to be manufactured by the same facilities normally dedicated to the production of electric power prime movers. This caused a virtual prohibition on the expansion of generating facilities for utility systems. It was the coordination of the large interconnected power systems that made possible the carrying of the war loads.

While each area did an outstanding job, I believe two in particular stood out; namely, the North Central area and the Pacific Northwest area. The former was close to being a 100% steam system, and the Northwest was an almost 100% hydro system.

The North Central area included Ohio, Indiana, West Virginia, and portions of Pennsylvania, Michigan, Kentucky, Tennessee, and Virginia, and some score of operating utility systems along with scores of municipals and cooperatives. This was a heavily industrialized area and therefore was particularly crucial to the successful prosecution of the war. Every effort was made by the group to maintain up-to-the-minute information on capacity and load, to schedule maintenance outages as if it were a single system, to pool reserves to the fullest extent, to make available spare equipment to one another—in effect to operate the area as a single system in so far as was possible.

Many steps were taken to conserve materials. Municipal operations, normally utilizing very small machines and thus uneconomical in their demands on material and labor on a per unit basis, were asked to purchase power rather than install new equipment. New transmission facilities were deferred if those of a neighbor could be used. As an example, the new Burger plant of the Ohio Public Service Company was about to come into operation. The company planned a 138-kv transmission line from the plant to the Ashland, Ohio, area, over 100 miles in length. However, an agreement was made whereby the transmission facilities of the AEP System were used during the war and the expenditure of this critical material saved for the war effort. A similar case occurred when the Columbus and Southern Ohio Electric Company agreed to transmit from one of its plants over facilities of AEP. There was probably no closer coordination at any location in the country than in this North Central area.

In the Pacific Northwest, war loads, especially aluminum reduction, were expanding at a rapid pace. The Northwest situation was further complicated by the public-private hassle that was at its height at this time. However, excellent coordination was achieved among the investor-owned, federal, municipal, and public utilities districts in the states of Washington, northern Oregon, Montana, and Utah. The systems included Bonneville, the large Seattle and Tacoma municipals, several investor-owned utilities, and many small municipals.

It was found that many transmission lines could carry substantially increased loading, and tests showed that the 109-mw Grand Coulee units could carry 128 mw without exceeding temperature limits. The step-up transformers at this plant were overloaded, but temperature was kept under control by water sprays. Extreme care was taken to coordinate the many reservoirs for maximum overall energy output. (The problem in the Northwest was energy, not peaking capability.) Changes were made in the operation of the Bonneville and Grand Coulee plants to give maximum energy output.

It was estimated that the power resources of this area were increased in excess of 200 mw by coordinated operation of reservoirs and maintenance, with additional savings from reduced reserve capacity requirements. The total area peak load at the time was 2630 mw.

Old steam plants were brought into operation, including two hogged fuel plants. Every effort was made to save strategic materials. One example is that relating to power supply to the Bremerton Navy Yard across the strait from Tacoma. Bonneville wanted to build a new line to serve the yard, but Bonneville and Tacoma were persuaded to pass this power through the Tacoma system. This presented some serious operating and service problems for Tacoma, but the plan was carried out and worked satisfactorily.

This coordinating group did an outstanding job, and all defense loads were carried continuously. A measure of this is the fact that at no time was it necessary to invoke the curtailment provisions of the OWU Limitation Order L-94 (a friendly action for legal purposes, which never had to be used).

The coordinating success story was the same for each area and it is no wonder that it led J. A. Krug, head of OWU and later of the WPB, to declare in a speech at Chattanooga that "power was never too little or too late."

P

Postwar Conversion Period

In some ways the postwar conversion period was even more difficult. The pent-up power requirements developed during the war and the general economic boom following the end of the war led to rapidly increased loads, but the utilities were behind on facilities because they were allowed only a minimum of new capacity during the war. However, the experience in coordination during the war made possible the mobilization of resources to meet the increased loads. The largest utility in the North Central system on the year's peak day in 1948 showed a reserve of 1 mw, but the load was met with no curtailment of loads.

T

The Korean Situation

Here, again, through the functioning of these coordinating areas, with the help of the Defense Electric Power Administration (DEPA), it was possible to meet the defense and civilian requirements with a minimum of strategic materials for new generation and transmission facilities.

T

The Supply of Interim Power to the AEC Oak Ridge Diffusion Plant in 1951

With the expansion of the diffusion plant at Oak Ridge and the new plant at Paducah, Kentucky, the TVA could not place new capacity in service fast enough to meet the AEC construction schedule and, as a matter of fact, could not fully meet its other loads. In December 1949 the surrounding utilities were asked to give assistance. Once again they coordinated their operations. During the period mid-1951 through 1954, the North Central system delivered some 5.6 billion kwh to TVA for its use and for delivery to the AEC. Deliveries from the Southwest power group to TVA amounted to 7.5 billion kwh and companies to the south furnished similar amounts.

T

The OVEC Story

When the war was successfully completed and it became evident that, because of the cold war, we had won the war but we had not won the peace, the concept that the atomic bomb could be enormously expanded in effectiveness by its further development into the hydrogen bomb gained strong support, as did the idea that what was considered initially as offering nothing more than a strategic weapon could be expanded into a tactical nuclear shield of enormous scope and effectiveness.

This, however, involved a huge expansion in the enrichment capability of the country not only at the original Oak Ridge site but at two new sites, each one of which was approximately as large as the expanded Oak Ridge project. One of these eventually was located at Paducah, Kentucky, and the other at Portsmouth, Ohio.

In Lecture IV of a book which I authored, which appeared in 1969, I described the project that was undertaken under the sponsorship of 15 electric utilities in the Ohio Valley to supply the full energy requirements

Table 3.4 Ohio Valley Electric Corporation: A Ten-Year Comparative Summary of Performance

		Total Power Delivered by OVEC to AEC	
Year	Demand (kw)	Total Energy Delivered (billions kwh)	Unit Cost—Basic Agreement (mill/kwh)
1957	2,137,000	17.064	3.985
1958	1,960,000	17.040	4.029
1959	1,958,000	17.060	4.041
1960	1,957,000	17.122	4.018
1961	1,910,000	16.642	4.008
1962	1,809,000	15.769	3.985
1963	1,807,500	15.760	3.981
1964	1,806,750	14.241	3.947
1965	1,206,750	10.510	3.971
1966	1,203,000	7.923	3.864
Arithmetic Mean			3.983

of the Portsmouth area diffusion plant.[5] This load eventually involved a demand in excess of 2 million kw of capacity and close to 18 billion kwh a year. Table 1 of that chapter (here reproduced as Table 3.4) gives a 10-year comparative summary of performance.[6]

Since then, information has become available on the cost performance of TVA, which supplied all the electric energy requirements at the Oak Ridge plant and also more than 50% at Paducah, and of OVEC. This is shown in Table 3.5.

[5] Philip Sporn, *Technology, Engineering, and Economics,* The M.I.T. Press, Cambridge, Mass., 1969.
[6] Ibid., p. 138.

Table 3.5 Cost of Electric Energy to the Three Diffusion Plants of AEC Including Cost of Replacements (mills per kwh)

	1964	1965	1966	1967	1968
OVEC Portsmouth	3.963	3.993	3.878	3.957	3.943
TVA Oak Ridge and Paducah	3.920	3.943	4.002	4.082	4.099

An examination of this table certainly gives no foundation for any claim that public power in this case has succeeded in taking care of the public interest in the case of a major national need any more successfully than private power has been able to take care of a similar requirement. As a matter of fact, the reverse may be true, particularly in view of the pending major rate increase for all TVA customers.

S

SUMMARY

Thus in summarizing the performance of the United States power system in the area of protecting the consumer interest, we come to an inescapable conclusion that compared with public systems in four major countries of the world—Russia, United Kingdom, France and Italy—its performance has been significantly superior from the standpoints of coverage and availability.

As for price, Table 3.3 shows an amazing 14-year record of consumer prices rising 30%, the wholesale commodity index rising 16.6%, while the average res-

Table 3.6

	Average Current Price for Industrial Electricity (mills per kilowatt hour)
Norway	3.29
Sweden	7.74
United States	9.08
France	13.54
United Kingdom	15.48
West Germany	16.8
American Electric Power System for 1969 [added by the author]	7.44

idential electric cost went down 18.5% and the total ultimate consumer price of electric energy distributed by the investor-owned utilities went down 11%. Add to this the tabulation that the chairman of the Swedish State Power Board prepared in discussing figures of cost of energy for industrial purposes in fiscal year 1970, which I presented in Lecture 2 and which I repeat here as Table 3.6.

It is obvious that, with the exception of Norway, which falls into a category of superior hydro resources that cannot be termed anything else but sui generis, the United States fully holds its own with Sweden—still predominantly a hydroelectric country of vast resources—even though Sweden's industrial rates ran close to 15% lower. As for France and the United Kingdom,

France's industrial energy rates are 49% higher, and the United Kingdom's are 71% higher than those of the United States.

As for the broad socioeconomic foundation for lower public power costs, I have gone to great lengths and believe I have shown conclusively that the only solid basis for the judgment that this is possible is if the cost of public power is lower. But the idea of lower costs is a mythical story, based on bad socioeconomic arithmetic, and there can be no differences in cost except as a result of more efficient organization and operation of public power enterprise.

So much for the consumer interest.

As for the public interest, it seems to me we have demonstrated clearly that in any challenging situation of public need in the United States private enterprise can be depended upon to, and did on numerous occasions, respond with the same degree of responsible action and concern for the public good as did public power, and both sectors found it entirely possible to cooperate completely in the best public interest.

The questions raised in this lecture, it will be noted, are not only important but difficult to answer. But they have been answered. Yet I do not believe we are ready to close our discussion, because electric power, an industry almost 90 years old, has most of its history ahead of it. The prospects of its quintupling in the next 30 years are clearly foreseen; in fact, it is almost impossible to arrest its course of development. It will, however, not happen easily. The many complex problems confronting the industry today will become much more complicated by virtue of this expansion. There is no question that the successful solution of most of these problems is going to have a heavy influence not only on the development of the industry but on the maintenance of proper service to the people of the country, the optimum utilization of electric energy, and the proper protection both of consumer and public interest. What these problems are, how the solution of these problems is going to be influenced by the kind of organization the industry will gravitate to, and, more importantly, what influence their solution should have on the kind of organization the country chooses, I hope to probe in Lecture 4. When I have finished that, I hope to be in a position to try to draw some conclusions.

4

Lecture Four

THE PROBLEMS AHEAD FOR THE ELECTRIC ENERGY SUPPLY SYSTEMS; SOME POSSIBLE DEVELOPMENTS IN THE UNITED STATES

The electric utility industry is beset today by many technological, financial, environmental, and public relations problems of great complexity and urgency for which proper solutions must be found. The difficulties stem from some major questions that have been asked about energy use on planet Earth, energy availability, energy conversion, energy transmission, and their social-environmental effects.

Because of these problems, the very ability of the industry to continue to supply adequate electricity for all the nation's needs is being brought into question. In the words of *The New York Times*:[1]

The power shortages bedeviling this city in recent days provide a reminder that adequate electricity for all communities and the nation's needs can no longer be taken for granted. Moreover, there is every reason to

[1] Editorial, "Power Shortage," 31 July 1970.

suppose that present problems will grow increasingly acute over the passing years as the population rises and as ever more difficult compromises must be sought between growing power demands and the environmental damage caused by hydrocarbon and nuclear fuels.

One of the strongest reasons for the almost certain increasing acuteness of these problems is that demands for additional electric energy in the next 30 years are expected to rise to a figure of 7200 billion kwh per year, the generation of which will require primary energy equivalent to 2125 million tons of bituminous coal. In the next 30 years, there will occur a fivefold increase in electric energy use and a corresponding increase in the primary energy required to supply the additional 1,335-000 mw of generating capacity that will have to be built. The problems to be faced will become more difficult because of the diversity of fuels that will theoretically be available: gas, oil, coal, and nuclear in both converter and breeder reactors. The difficulties will become aggravated because 300 new generation sites will have to be found and developed as loci for generating plants of an average size of 4500 mw each. Even after the sites have been found and the plants built, it will require the utmost in vision, daring, and inventiveness to build the necessary interconnection transmission in voltages up to possibly 1500-kv alternating current and at no lesser levels direct current. Further, all the new and more complex system facilities involved in this expansion will raise problems in pollution of air, water, space, and in aesthetics. Along with growth in per capita energy use and the expansion of the percentage of total energy that will be electrical (to a point where 50% of the total energy used in the year 2000 will be in the electric form) and with the growth of the ramifications that can be expected to develop as energy and electric energy use expand, the patience of the public and communities with fading voltage and unreliable supplies will diminish. Insistence on a completely and fully reliable service will become an implacable demand.

F

FUTURE PROBLEMS

Let us take a closer look at some of these problems so that we may be able to judge better the importance of their solution and the likelihood of that solution being assured to a greater or lesser extent by either public or private power. I want to examine seven of the most important or critical problems.

F

Fuels

By the end of 1970, the total electric capacity supplied by nuclear fuel will be less than 10,000 mw or less than 3% of the total. By the year 2000, the nuclear capacity is expected to be 600,000 mw, and it is expected that it will produce 50% of the total of 7200 million kwh. Of the remaining 50%, 35% would be supplied by coal.[2] Thus of the total of 2125 million tons of bituminous coal equivalent of primary energy, 745 million tons would be coal. This is more than double the 1970 estimated burn of 325 million tons.

The utilities of the country are in trouble on coal. Recently, *The Wall Street Journal* carried a story headlined, "TVA Says Coal Supply May Run Out, Causing a Shortage of Power," [3] and another, "American Electric Says Its Supplies of Coal are 'Dangerously Low.'" [4] Thus the largest publicly owned electric utility and the largest investor-owned utility in the country were, apparently, in grave danger of an interruption of their power supply as a result of a shortage of coal. Scores of other utilities are in even worse shape.

Some of these troubles are due to the reaction of the coal industry to the scary doomsday projections for coal that the early enthusiasts of atomic power issued. Some of the difficulties are the result of the recent ownership monopolization of a good part of the coal industry by the oil industry. Some of the troubles stem from the irresponsibility of many coal operators and coal miners. And some are the result of our lack of a national energy policy. At the national level we have never seen the urgency of adopting, nor the development of a possible energy crisis in not adopting, a national energy policy and setting up a high-level commission or similar body to deal with the problems re-

[2] Philip Sporn, "Developments in Nuclear Power Economics, January 1968–December 1969," Hearings before the Joint Committee on Atomic Energy, Congress of the United States, U.S. Government Printing Office (Washington, D.C.: 18–20 November 1969).

[3] 20 August 1970.

[4] 4 September 1970.

lated to the implementation of such a policy. However, much of the difficulty is caused by the failure of many members of the public utility industry itself, both private and public, to anticipate the difficulties that are pressing so heavily on them now and to take remedial steps early enough.

The August 20 story in *The Wall Street Journal,* previously referred to, quotes a spokesman for TVA as saying, "The basic problem, to sum it up, is that coal production isn't being expanded enough in relation to demand." Why, one can properly ask, did not TVA provide for a 30-year supply of coal for each unit or plant it built before ground was broken for the first boiler foundation? After all, what is more useless than a coal-burning plant without coal to burn? Keeping in mind that the development of a new mine calls for coal reserves, management skill, capital, and three to four years, if new mines had to be opened to bring about the necessary coal supply, was it not the responsibility of the power system management to carry out or to arrange for the carrying out of all the necessary programs in advance of the building of the power plant? Was not assuring that coal production was being expanded enough to meet its needs and having that capacity contractually committed to its system an indispensable part of a comprehensive system planning and development operation?

The impending crisis in coal supply at the country's power plants in general (to prevent which the imposition of severe restrictions in the export of United States coal is being asked) leads one to wonder whether the complaining utilities were ready to order coal at the time the coal operators needed some business very badly to keep afloat. In a private enterprise economy, whose responsibility is it to have available the coal necessary to forestall serious shortages of electric power? Whose responsibility is it that General Motors, Ford, and Chrysler have the steel to manufacture all the cars their customers want to buy?

The uranium supply and cost of future cores will also present major problems. The entrance of the oil industry into both coal and atomic fuel has established a near monopoly position in the three important competing fuels that enter into electric energy, and this will aggravate pricing. This fuel situation poses a great threat to the utility industry; the industry should have done something about averting the problem when it first appeared on the horizon.

Atomic fuel is not quite ready to be converted into electric energy after it is mined and converted into yellowcake. First it has to be enriched. Looming on the horizon is the development of a messy situation in atomic fuel enrichment should the federal government, which now owns and operates all the enrichment facilities, decide to leave that business. Here, I believe, the obvious solution is the assumption by the utility industry, acting as a whole, of the responsibility for the enrichment operation. Up to now, this has remained a neglected opportunity to assure a reliable and reasonably priced supply of fuel for all atomic plants.

D

Development of Larger Generating Units and More Intensive Development of Sites

The leading electric utilities today are exploiting 800 mw on a single shaft and 1300 mw on two-shaft machines, the so-called cross-compound machines. But experience at 800 mw with the product of at least one of our two leading manufacturers[5] raises grave question as to whether that manufacturer has mastered the technology of building a large single-shaft machine. While this can be avoided by going to the cross-compound design, the economic price is stiff. The need to obtain greater economies and to overcome inflationary effects impels action by management of quality to try the single-shaft design not only at 800 mw but to develop it over a period of time to eventual ratings as high as 1670 mw. This last rating offers the possibility of building a 5000-mw plant with only three units. Even over a period of time, this is a major undertaking.

In their haste to develop shop capability of filling demands by the utilities for generation equipment, some of our manufacturing organizations have neglected research, development, and quality control. This neglect has had most telling effect on reliability. A single issue of an industry newsletter details accounts of five major cases of breakdown of generating equipment that have led or may lead to court action, indicating that being

[5] Special report, "Turbine Failures Raise Big Questions for Utility Industry," *Electrical World Week,* 31 August 1970.

promised a delivery and even erecting a unit and turning on steam as scheduled are not enough if the machine cannot stay on the line after it has been synchronized. Thus, buying and installing generating capacity and sponsoring and accepting new designs of advanced technology is something more complicated than giving a manufacturer an order and writing a contract.

It may be that these particular suits are merely an indication of the need for a more "abrasive buying" attitude by the utilities. However, it is quite clear that reliability and availability are going to represent a major problem in the future. It is also obvious that the utilities will need to explore and evaluate all new developments so as to be able to judge whether the promises or offered contractual commitments of the manufacturers are likely to be fulfilled. A preliminary question that the utilities will need to ask and answer is whether their engineering organizations have the assured competence to explore and judge such developments. This becomes a complicated problem.

H

Higher Voltages

For many years a quite pervasive myth had developed in too many electric utility circles that higher transmission voltages were needed to transport large blocks of electric power only over long distances. Since many systems were circumscribed in the area of service, they were quite sure that they had no need to worry about transmission. This is illustrated by one system that, having developed

120 kv as a system voltage as far back as 1919, did nothing further for the next 45 years, even though its demand or energy generation had in this interval grown substantially 20-fold. It thus grew in size as a utility but never developed into a system.

But the grave danger of that poor thinking has not been entirely removed. At the present time we have well established in the United States alternating-current transmission at 345 and 500 kv. The basic research and developmental work in the range of 230 and 500 kv had been carried out in the 1945–1950 Tidd research project sponsored jointly by the AEP and Westinghouse engineering organizations.

Because the system had quadrupled in size since the end of the war and given every promise that it would double, redouble, and redouble again in the next 30 to 35 years, about six years ago it became evident to a number of people on the AEP System that 345 kv, which had been brought into being in the middle 1950s, would not mark the endpoint in the system's requirements of higher transmission voltages. The need to exploit larger size units, to exploit each site more fully, to cut the capital and operating costs of units and plants, to obtain the higher thermal efficiencies incident to larger sizes of units and plants, to protect the outage of single units and of whole plants, the consequent need to increase materially the capacity of regional internal and external interconnections—all these confirmed previous judgments that essentially system capacity and transmission capability had to march hand in hand. Higher voltages were needed as systems grew in size.

A research program that had been undertaken on voltages beyond 500 kv was then speeded up, and two years later a decision was reached to proceed with a major program of 765-kv transmission, the first part of which has recently gone into service.

If the principles and ideas I have developed here are sound, namely, that the need for high voltage arises from growth in size of system not from lengthening of distance of required transmission, then obviously, since we can as yet not see the end of growth, 765 kv itself is not the end point. Before higher voltages or the ultimate higher voltage for overhead construction is decided upon, a great deal of research and economic study need to be carried out to determine what this voltage or voltages should be to solve some of the difficult aspects of overvoltage control and of line insulation and insulator performance and, particularly, to be sure that all the possible environmental polluting aspects have been completely studied and researched and appropriate solutions found.

While all this needs to be done and explored thoroughly, these studies should not be used as a basis for delay. There are plenty of interconnections that can be brought into being constructively at voltages up to and including 765 kv. There has been much too much tendency in the past in planning and implementing plans for interconnection transmission again to wait and to postpone; nothing but great harm to the performance of power systems can result from such a policy.

Here again, great organizational ability and pioneering courage will be required to assure implementation of proper policies.

S

System Reliability

The effects of the traumatic blackout experience in the Northeast on November 9–10, 1965, have not yet been eradicated, if eradication includes not only the realization of the cause of the failure but decision to remove the cause and bring the likelihood of a repetition into the remotest of occurrences. A good deal of work has been done. Many hearings have been held by commissions and congressional groups; elaborate computer studies have been made by competent people; many trips to foreign countries—England, France, Italy, Austria—have been made and advice solicited from people, who are sometimes competent to give advice and quite frequently incompetent; reports have been written; bills have been introduced into legislatures; but the problem of assuring reliability and eradicating the likelihood of a recurrence of a major system power failure has certainly not yet been solved. Yet the basic requirements of a solution are simple.

It must be recognized that one cannot have a system merely by building generating capacity. The mechanism for transforming a series of sources of power and areas of load into a power system is the interconnecting transmission network. This network is not a static affair. It cannot be built and forgotten. It has dynamic characteristics, and it changes with the growth of load and generation. Therefore, the network has to be continually studied and any deficiencies scheduled for quick remedial action. What this action should be is frequently complex, but it can be determined quite precisely; while some judgment is required in its implementation, the leeway is relatively small.)

A system composed of many power groups with separate managements and ownerships, operating as a closely knit grid, introduces one difficulty, that is, the surrendering of sovereignty. It is very hard for any number of people to accept the idea that some group, even of highly competent members, of the interconnected companies will study a specific company's plans and tell that company that a line it plans to run from *A* to *B* will perform much better as a part of the interconnected system if its route is changed from *B* to *C*. Even though it is absolutely obvious that such coordination of system development is indispensable to assured system reliability, I have had technical heads of such systems tell me that before they accede to anything of this kind they would wait to get an order from some federal commission. But this is merely bad judgment, aggravated by irritation with the facts of life that make such a situation necessary. It does not make any sense to refuse to accept the judgment of a competent group that has no special interests of any kind to serve and whose sole dedication is to reliability.

This problem will obviously become tougher to handle when the power systems of the year 2000 come into being. Then, as now, the human tendencies will be displayed that stem from fragmented interests, for example, shortchanging efforts of plant expansion in the interest of protecting earnings of a particular member of the group against such expansion in the interest of protecting the service of the entire area. It can be fatal in communications, as witness the breakdown in the quality of telephone service during the past 2 years in New York, and it can be equally fatal in a power situation.

Thus there will be required not only complex transmission facilities but sophisticated and continuous thorough analysis (made in advance with a view of prophylaxis and not after failure in the form of a post mortem) of system performance and a dedication to the principle not to let anything interfere with building solidly. For achievement it may require some strong central authority from within the industry members of each area or region. It will also require vision, a sense of responsibility, dedication, skill in the form of highly trained and competent personnel, and resources. I have discussed this in greater detail elsewhere.[6]

E

Environmental Pollution

Environment and its pollution have now become household words. The situation has long since passed beyond what it was only a short while ago, that is, a characteristic of Los Angeles which as a result of special, unfor-

[6] Philip Sporn, *Research in Electric Power* (New York: Pergamon Press, 1966).

tunate meteorological conditions suffered from smog. Nowadays we hear and see daily in the press, in radio, on television, and in the halls of Congress, denunciations of all kinds of pollution—soil, water, air, upper atmosphere, noise, aesthetics—as a result of burning fuels, generating electric power, carrying on chemical and metallurgical operations, running of automobiles, and the building of high-voltage transmission lines. The people responsible for generation, transmission, and distribution of electric energy have been a particularly prominent target for attack. It has in consequence materially interfered with the business of supplying and operating facilities to furnish electric energy, and it has become a nightmare to numerous electric utility managements all over the country.

The blame is not confined to managements specifically; it is more broadly allocated to them generically, as exploiters or as applicants of technology, for profit in the case of the private sector, or merely carelessly without concern for the public weal in the case of the public sector.

This souring on technology may be paradoxical because the world we live in has come into being as a result of technology. It is no accident that between 1915 and 1967 our population doubled from 99 million to 200 million, while the number of people employed in agriculture declined 64% from 13.6 million to 4.9 million; it is the result of technology.

Technology, which made possible our increased productivity, has given us a higher level of economic welfare for the average man than that produced by any other civilization. The increased production and productivity that we will need in the future to take care of our greater population can come only from further improvements in technology stemming from more efficient and more automated processes, better management, and the greater availability of inanimate energy to every worker. This production organization has been something that the underdeveloped and developing nations have been exerting themselves to emulate.

However, there is a negative aspect to this. While all this was taking place, there has been an amazing growth of concern about the negative effects of technology. This has reached flood proportions, particularly regarding the pollution of the environment. We are worried about the potential damage to the ecology of the planet and in the words of Emmanuel G. Mesthene[7] about "occupational and social dislocations, threats to the privacy and political significance of the individuals, social and psychological malaise."

Of course, we cannot give up the continuing pursuit of technology. The casualties of such a withdrawal from technology would, according to Max Ways,[8] "be heavier than many suppose and the retrogression would have to be deep, it would have to go back a long, long way." Again in the words of Ways:

A century ago we had already slaughtered the bison, felled the eastern forest, and degraded the colonial cities. Retreat to the 1870 level of technology, while not giving long-range protection to the environment, would place

[7] Emmanuel G. Mesthene, *Technology and Culture, 10,* no. 4.

[8] Max Ways, "How to Think about the Environment," *Fortune,* February 1970.

the median American standard of living far below the 1970 poverty line. Among the consequences of such a retreat would be the closing of 75 percent of the present colleges and most of the high schools. We would give up not only autmobiles and airplanes but also mass education and social services. Grandpa would return to living in the abandoned hencoop.

And yet, so well balanced a newspaper as *The Wall Street Journal,* in deprecating the inadequacy of long-range (reaching back at least 20 to 30 years) planning methods that led to the need for the proposal to add 1600 mw of fossil-fuel capacity to the Astoria generating plant, concludes with this observation:[9]

> The episode suggests to us, however, that if we as a society are to make any progress whatever with air pollution, we will have to readjust the unconsidered assumption that when we face a choice we will in every instance and as a matter of course choose needed power over needed improvement in the air. New York has seen power cutbacks during the peak periods of the hottest days this summer, with hall lights turned out, subways moving slower and television pictures smaller. All things considered, perhaps this is not after all the worst thing we could choose.

I myself believe in the feasibility of obtaining both increased energy for man and environmental protection and that, although it may be difficult, the two are or can be made compatible.[10] However, it is necessary to recognize that possible environmental regulations significantly delimiting the development of energy supplies and of electric energy present a serious danger.

[9] Editorial, "Power and Pollution," 21 August 1970.
[10] Philip Sporn, Editorial, "Energy for Man and Environmental Protection" *Science,* 31 October 1969.

This possible limitation obviously is a much greater danger for the future. The environmental problem touches an instinctive raw spot in every human breast—a deep concern for the safety of one's self, one's offspring, and the very survival of the human race. Its ramifications will, in the atmosphere of the coming quintupling of electric energy output, call for the greatest foresight, vision, understanding, and courage to act to forestall a crisis and to avoid being confronted by the need of choosing between meeting demands for additional service and environmental safety. This is no longer easy today. It will be more difficult 15 to 30 years from now.

C

Costs of Electric Energy

It was only a few years ago that a school of utility analysts appeared who out of their deep wisdom and advanced studies came to the conclusion that the electric utilities of the country had ceased to be attractive loci for investment capital because of the decline of the rate base. In nontechnical language they meant that the utilities had ceased growing, hence earnings too would stop growing if they had not already done so. The amazing response of a considerable group of utility executives could have been predicted. After all, they reasoned, rate base originates in buying equipment and building electric property. It can be increased by building more offices, public information centers, storerooms, garages—and there is always need of these. It can be further increased by building transmission lines on rights-of-way held in fee rather than controlled by permanent easement. Generating reserves can be increased and many more things of the same nature can be done. They proceeded to do many or all of these.

Both the analysts and the executives were wrong in their conclusions, and their policies were based on fallacy. They forgot to study the history of growth of electric power and the part in it that continual lowering of prices played.

One of the important factors contributing to the expansion of electric use has been the decline in price of the product made possible by the industry's continuous search for, and effective deployment of, technological progress. This feedback relationship, with lower price making possible additional expansion utilizing improved technology and new technology making possible still further reduction in price, has been and will continue to be an essential—indeed an indispensable—input factor to sustain the dynamic growth of electric energy consumption.

Technology has been the key to growth. Technology and growth have moved together, and they have been closely interrelated. In the main it was technological advances that brought about rapid growth, but, in reality, this has been a closed cycle. The rapid growth also made possible many of the technological advances. Particularly has this been so in the last quarter-century —at least up to a few years ago—when rapidly growing use of electric energy and substantial technological advances made possible a decline in the average price per kilowatt hour despite major increases in the general price level.

During the last 40 years, electric utility generation increased almost 20-fold while gross national product less than quadrupled. This has involved a change in the character of a considerable part of the electric utility industry from relatively small, more or less isolated, systems into much larger and more efficient integrated systems. Many of these larger systems in turn have interconnected with their neighbors and thus have obtained further advantages in exchange of reserves during maintenance, emergency help, staggered construction, and economy and seasonal energy exchanges. Some efforts have also been made to establish major pools to exploit the advantages of larger scale construction of generating capacity. A number of such pools that were brought into being have recently separated; on the whole, the pool idea has not turned out to be a particularly successful one.

But things have changed. Between inflation, heavy increases in the price of fossil fuels, increases in the cost of atomic power to 7 mills per kwh at the switchboard at 75% capacity factor, soft buying and building, we have now reached a stage where major increases in rates have been received or applied for. But higher costs and higher rates, even where they have to be accepted, are not a proper answer.

As I pointed out in a report to the Joint Committee on Atomic Energy:[11]

Seven mill nuclear power at the switchboard at

[11] Philip Sporn, "Developments in Nuclear Power Economics 1968–69."

75% capacity factor is simply not good enough to heat water, to heat houses, to make electric steel or ammonia or chlorine, to reduce alumina to aluminum, to smelt ferro-alloys, to desalt sea water, and to convert to electricity the many other energy operations our society needs to have done so as to eliminate environmental pollution. By electrifying we make possible elimination of combustion of fuels except in large-scale, highly regulated and controlled fossil-fueled operations. Of course, atomic reactors take care of this matter even better.

Is there then a better way? The answer is yes, by resisting at every point where higher costs enter the system. Are equipment costs too high? Assuming specifications are not burdensome, try more abrasive buying. Or try foreign competition. There are today foreign manufacturers that make steam-electric turbines that are as good, transformers that are as good, switchgear that is as good, reactors that are as good as the best American product and in some cases better than our best. Where prices are 20% to 35% less, there is every incentive, not only from the standpoint of the buyer's interest but also from the standpoint of the national interest, to encourage the foreign manufacturers so as to give our own time to sharpen their technical wits so they can live in competition with those who have succeeded in doing a better job.

Are labor costs too high? Carry through an intensive study on how to make a man-hour of labor do what $1\frac{1}{2}$ or 2 hours have done before. There are still plenty of locations where that is possible.

Are fuel costs too high? Has fuel possibly reached an unobtainable position? Why not develop your own supply as Duke Power has just announced?[12] Incidentally, the solution for its coal problem chosen by Duke is no startling innovation. Other utility systems, including the AEP System, have for many years been mining significant percentages of their coal requirements.

The exploration of all the avenues open and the exploitation of all the potential mechanisms for cost reduction are not going to be easy, simple, one-time operations; it will take a great deal of skill, understanding, planning, and hard work.

R

Research

Research deserves to be a separate and final item in the discussion of the problems that electric energy supply systems are coming into. Throughout this lecture I have indicated the need for continuous research, and this is certainly going to be an indispensable part of the solution of many of these difficult situations that are going to develop in the next 30 years. But research will or should enter to a much greater degree in utility management planning, building, and operation than it has in the past.

The principal reason for the judgment that more research will be needed exists in the dynamic character of electric power use. As I have already pointed out,

[12] "Duke Power Purchases Two Coal Mines: Negotiating for Another to Guarantee Coal Supplies for its Generating Stations," *National Coal Policy Conference Newsletter,* 3 September 1970.

the United States will have in the next 30 years not less than a quintupling of electric energy use in terms of the 1969 base. It is simply inconceivable that present-day methods of gathering the primary energy involved, converting it into electric energy, transporting it from the point of conversion to the distribution centers, and then distributing and utilizing it are going to continue to be good enough for the next 30 years. Within that brief period, facilities for production and transmission and devices for utilization four times as big as those produced within the last 88 years will have to be created. Contemplating this, one can conclude safely that a decision to carry out much expanded research will be inevitable.

The question is how will this research be carried out. An effort is now being made to do this by a centralized group of all utilities operating with the title of National Research Council. But any sober appraisal of the growth that lies ahead and the research challenge that is presented by that growth yields the inescapable conclusion that leaving the research of a great industry —the most capital-intensive industry in the country— to a single body is a rash and risky thing to do. A single body simply cannot provide the mechanism for sensing areas of research, reaching decisions regarding the need for work in those areas, and carrying out effective measures for implementing research programs. The United States is much too vast a country, too diversified in its economy, and too regionalized to make a single center for research fruitful or to make such a center anything but a device for discouraging research.

One is therefore forced to the conclusion that the basic mechanism for carrying out research in power in the United States will have to be in regional organizations. This makes good sense. For example, in generation, the availability of local fuels, that is, fuels endemic to the region, is bound to play a significant part. In transmission, local geography and topography will play critical roles in design, construction, and operation of systems. In distribution, topographical, meteorological, and sociological considerations will surely play a significant role in the basic design adopted and in the materials selected. The final argument for the regionalization of research is that it gives much greater opportunity for diversity of thought and greater exercise of individuality and responsibility in the conduct and control of research than would complete centralization.

It is obvious that adopting a regional program is bound to introduce some adverse characteristics into a national research program. For one thing, there may be some overlapping. Provided this is not too extensive, that characteristic, far from being objectionable, is one to be sought. But there may also be projects that spread so completely across all the regional areas and involve so large an expenditure that anything but a single centralized program would make no sense. In those special cases centralization can be accepted, but it would be controlled by the area groups and thus centralization would not get out of hand. In short, what is being said here is that it would be essentially a regionalized program with some compromise, that is, recognition of the propriety of some centralized research.

How much does this regional approach differ from what we have at the present time or toward which we have made a start? The answer is, almost completely. It is different because the essential emphasis will be on a regional setup. The current National Research Council program is a voluntary one and therefore offers too great a temptation to leave the work to be done by the enthusiasts for research and let the others who consider themselves too smart to waste money on research to gather the benefits of it without expense. Since this has in it no element of fairness, one is forced to the conclusion that the only way of financing such an operation is by an assessment, and the fairest arrangement is one that is based on gross electric revenue.

This is not the place to attempt to develop in detail the organization of such a program, the lines of demarcation between regional and centralized research, the governance of the various groups, and the means of exercising control. All of these agencies and mechanisms need to be carefully planned and critically discussed. Some legislation may be required. Also, control will be very important.

It will be essential to set up a body to function as a kind of father confessor, special pleader for, and special forum (holding of hearings) to develop and record research needs, to receive reports on work in progress, and projections for the future—to act, in short, as a high-grade gadfly and as a general stirrer upper of the question of research in power in its manifold implications.

The carrying out of a research program on such a scale is so different from the situation throughout the history of the industry that one naturally asks, why make a change? Hasn't what we have done in the past been good enough? Here, of course, the answer has to be based entirely on judgment and faith and evaluation of what is ahead. After cogitating on the great expansion ahead and on the many difficult problems that this growth will raise, I believe that there can be no other conclusion but that we will need—indeed, we shall have to have—something better, much better, than we have had.

A
APPRAISAL OF FUTURE PERFORMANCE

I have come to the end of the presentation of the seven critical problems, the solution of which ought to be heavily influential in our choice of organization of the much greater power industry that is destined to come into being. Difficult as it will be, the appraisal of the likely performance in the future, which I shall next consider, may, unfortunately, be meaningless, because we are currently suffering from confusion and serious delays in implementing utility decisions because of the public involvement in these decisions. It is unfortunate that the utility industry, both public and private, has not found the secret to getting public acceptance of utility decision making.

Some of the cause of this may be a variation of Hamlet's concern when, dying, he pleads with Horatio: "Thou livest; report me and my cause aright." Undoubtedly the utilities have not done a good enough job in presenting their cause. But the mechanism for reporting, presenting, and resolving their cause is sadly inadequate. It is quite evident that we must find better mechanisms than the public hearings we now employ on fragmented aspects of utility planning. Unless we do, we shall find the so-called public interest groups making it more, rather than less, difficult to resolve the seven critical problems in electric power I have just discussed, regardless of whether the facilities are privately or publicly owned. Even worse, they may be resolved to an antipublic interest end point.

But with moderate confidence that a mechanism for improvement in public acceptance of public utility decisions will be developed, let us move along to a final appraisal and evaluation of the likely performances of power systems in the future under the two forms of organization.

The history of public power in the United States shows few records that it originated in an ideological search for a better, socially more responsible, mechanism for handling that beautiful socioeconomic tool of an adequate, reliable, and economical electric power supply. In the municipal area its origin was political, aided by bad evaluation by the private utilities of the quid pro quo in the matter of a utility's duty to a community for the privilege of doing business within its boundaries. The Rural Electrification Administration (REA) was born out of desperation on the part of our rural citizenship in their efforts to get for their farms and homes the help of the electric servant. Certainly in more nationally oriented (hydro) projects, bringing the project into being as a step to get on to the ladder of political advancement was no rare exception.

In foreign countries public power was either as a result of political dogma (Russia, Italy, France after the Resistance) or doctrinaire ideology (as in England, where it was compounded by the bad results obtained by the all-pervasive municipal operations).

As far as the difference in organization influencing technical competence, there is a good deal to warrant the conclusion that private enterprise offers the advantage of single, clear, and unchanging primary objective; it permits management to exercise much greater freedom to optimize its behavior in pursuit of that objective. There is no evidence that public power has done a superior job in extending service or in the quality of service rendered. Indeed there is a recent regrettable parallel of management under both forms failing in its responsibility for protecting service, as witness the fuel crisis confronting the utilities across the country.

When we come to the protection of the public and the national interest, particularly in times of emergency, we find that the private sector has at least responded as fully as has the public. As a matter of fact, the response of both segments of the power industry during World War II was exceptionally and completely satisfying including the wholehearted cooperation between the two sectors at their points of contact. In the post-war period, OVEC has clearly demonstrated that, given the proper organization, private power can do as well in supplying major governmental requirements as can public power —better, in fact.

The mythical advantages of price under public power have been shown as apparent savings as a result of avoidance of socioeconomic responsibility and bad socioeconomic bookkeeping.

These valuations are important for their contribution to the passing of judgment, but they are nowhere near conclusive. More important than ideology or past performance are the likely performances in the handling of the series of great and difficult problems, which I last discussed, that will confront the electric power industry in the coming 30 years. Here is my judgment of the most likely respective performances in the solution of each of the seven problems.

C

Coal and Other Fuels

If you are concerned with fossil fuels, then it appears almost certain that possible investments running to hundreds of millions of dollars in fuel facilities will be called for on the part of the utilities. I believe this can be handled much more easily under private operation. The private sector has an on-the-whole successful record going back more than 50 years in the mining of coal by many organizations, such as Louisville Gas and Electric Company, Ohio Power Company, Appalachian Power

Company, Southern Power System, Allegheny Power System, and the list does not end here.

For some, but not all, of the same reasons this would be equally applicable to mining, processing, and the enrichment of uranium.

D

Development of Larger Units and Fuller Exploitation of Plant Sites

Where performance of system and choice of equipment can play so great a part in profitability, there is always the danger that the publicly owned system will act more rashly than will the private organization. With the utmost of goodwill toward the electric power operation in England and a high regard for the competence of its managements, one cannot but have grave doubt whether, if commercial considerations had been given enough weight, the British power system would have made such a heavy commitment in its early atomic power installations. Or negatively, if commercial considerations had been given sufficient weight in France, whether the ship of French power supply would have found itself mired in the nuclear bog that it has been in these many years.

H

Higher Voltages

While Bonneville has done some courageous things in the field of extra-high-voltage transmission, they were confined almost exclusively to the Northwest direct-current intertie, and undoubtedly political motivations entered to a great degree. *Electrical World* of September 1, 1970, has an interesting news item bearing on this observation:

Interior Dept. did not postpone federal 750-kv dc line between Celilo, Ore., and Mead Substation, Ariz., on time when it became apparent that the line was not economical, a General Accounting Office report charges. The report said information became available "as early as 1966 that the project might not be self-liquidating within 50 years." Yet in Jan., 1969, Interior told GAO the project would be fully used on a timely basis, provided that firm and realistic completion schedules could be maintained. It was not until May, 1969, that Interior announced the request for funds was being postponed and a contract for ac-dc conversion facilities at Mead Substation was being terminated. The report said $37 million had been spent, of which $10.4 million—plus $3.8 million in progress payments and an option on a second converter at Celilo—may have no value to taxpayers.

Many, many years ago Thomas A. Edison made the observation that technology and politics did not mix well. The validity of this dictum has not been attenuated by time.

S

System Reliability

The bringing together of many smaller systems and creating the basic condition prerequisite to attaining system reliability by having these groups surrender complete sovereignty over their facilities having an impact on reliability will be much more difficult with public than private bodies. Freedom of decision and freedom of action become more circumscribed when ownership is public.

E

Environmental Pollution

Experience in the environmental situation has not borne out the expectation of leadership from the public sector. I base my judgment on my professional experience and the experience of my family in carrying on a losing fight for many years against the coal dust pollution of our apartment in the city of New York located two blocks north and two and a half blocks east of a city-owned and operated steam-electric plant. I am not sure that I would subscribe completely to the criticism of one of the distinguished English authors when he said:[13]

> Britain has been spared strip mining and auger mining for coal, but the hills of Kentucky are being scarred by this utterly ruthless and extractive process. Downstream the good arable lands along the Ohio River are being spoilt by the silt blocking the waterway. Here, too, is another dichotomy, for the large consumer of this coal is that environmentally conscious body, the Tennessee Valley Authority. Should a state boundary act as a blindfold to what is happening on the other side of the watershed?

However, this charge leveled against TVA is not entirely groundless.

[13] Frank Fraser Darling, "The Reith Lectures 1969," (London: British Broadcasting Corporation, 1970).

On the whole, my judgment is that, with all the concern for profit inherent in private ownership, the proper balancing of costs versus environmental improvement and the many accommodations that will be required between municipal, state, and federal bodies and the utilities in solving the future environmental problems are more likely to be satisfactorily worked out under private than under public ownership of the utility systems.

C

Costs of Electric Energy

This brings us to costs. Electric energy costs, I pointed out, need to be lowered not raised. They apparently can be lowered by public ownership and operation. Is this a solid reason for expanding public power? I do not believe so.

It is my conviction that these apparent reductions are antisocial and antipublic interest. Again, take TVA as an example. TVA has been operating efficiently, but it has perhaps done so at a sacrifice of its own position as an important corporate citizen. From its total sales in 1969 of 86.4 billion kwh, TVA received an average realization of 4.5 mills per kwh. The payments in lieu of taxes that it made were $14.5 million, or 0.168 mills per kwh, 3.75% of its gross. This offers an interesting comparison point. The private utilities had gross sales in 1968 of 985 billion kwh, for which they received an average realization of 1.61¢ per kwh, and the taxes paid to federal, state, and local bodies were $3484 million.

This represents 3.5 mills per kwh, or 22.0% of their gross. Had TVA paid a similar percentage of its gross per kilowatt hour in taxes, its bill would have been approximately $85.5 million. The difference between this figure and what it actually paid was $71.0 million, and this, for all practical purposes, represented money not contributed to the support of the economy at federal, state, and local levels. This is roughly the amount by which TVA's customers were subsidized in not contributing through their electric bills to the support of the economy.

The English, who have struggled with this question, recently published an extensive summary on efficiency and costs in the public sector of their economy, prepared by George Cattell, formerly director of the manpower and productivity service of the department of employment, in which he says:[14]

> After completing two years' secondment from the motor industry to Whitehall as Director of the D.E.P.'s Manpower and Productivity Service, I now compare the efficiency, economics and performance of the agricultural industry with those of the private sector of manufacturing industry and the nationalized corporations. As the representative of the farmers I, too, am concerned that costs of essential services provided by the state should not be inflated because inefficiency is tolerated.
>
> My first inescapable conclusion is that had home food production been nationalized with coal, gas, electricity, postal services and so on, we should now be paying several times the current prices for, say, our milk, meat and potatoes. Farmers would now be Area Land Managers with salaries far above their present incomes, the National Farmers' Union would be extinct and the

[14] *The Times* (London), 28 August 1970.

elegant Agriculture House in Knightsbridge would be occupied by the National Food Board. . . .

Political doctrines aside, there is no case for extending the public ownership of industry. But the problems of denationalizing the industries now owned by the state are probably too formidable to be contemplated. The alternative is to convince their managements that they can and must live with increasing wage demands and stable prices for their own products or services. The only way I know of doing that is to subject them to continuous public scrutiny and show them in hard practical terms how they can cut their costs.

S

SUMMARY

Thus, I have weighed the pros and cons of private and public power in the United States and other countries. It is fair to say that from this I could say that in the United States, at any rate, the balance is in favor of private enterprise because of continuity of management, freedom to make long-term commitments, incentive to stay efficient, freedom to carry out technologic innovations, ability to attract the quality of personnel necessary, ability to plan ahead and bring about deployment of one's plans, and because of the greater likelihood of being able to solve satisfactorily, or solve more satisfactorily, the series of problems that will confront the electric power industry in the next 30 years.

Making this finding, I admit, troubles me. Private enterprise, which in my judgment has so many ideological advantages, possesses a considerable number of basic weaknesses. To mention half a dozen: the tendency for managements to grow, to put it crudely, fat and lazy; the propensity, and indeed common practice, of electing directors who really do not direct and only awaken as in the case of the Penn Central to that fact after tragedy; the tendency to play safe; the lack of imagination; and the lack of broad social understanding or sympathy.

There are some consoling offsetting facts: Public power top management, with perhaps almost as many cards in its favor ideologically, also has deeply imbedded weaknesses. To mention a few: excluding themselves from support of social activities of the nation; lack of control by anybody to check their judgment and to find whether, acting as they supposedly do in the public interest, their actions are actually designed to achieve that end; their tendency to feel smug and complacent on the theory of "where else can they go?" if people are dissatisfied with the job they are doing; the lack of incentive to stay efficient; with all their avowed intentions of acting in the public interest, their failure to assert leadership in the solution of socioeconomic problems; and, finally, their tendency to permit themselves to become politically dominated.

Contemplating these deficiencies in public power ought to reassure one as to the future of private power. Yet, looking toward the future, I continue to be deeply troubled. It is quite clear that the future demands our society is going to make on its energy supplies (more and more this will become synonymous with electric energy) will become heavier and more difficult. Technological problems will be more challenging. Environmental problems will become heavily loaded with conflicting criteria, each one apparently as important, if not more, than another. Because of the complete ramification of energy supply into almost every activity ranging from national defense to systems of control of atmosphere, weather, that is, sunshine, rain, and snow, and involving perhaps artificially created hurricanes as instruments of irrigation and crop control, absolute adequacy, continuity, and reliability will become a sine qua non. Always present will be the problem of costs.

How shall we influence the growth and development of what will become the biggest industrial organization in our society? Shall we let nature take its course? Can we as a matter of fact do anything else? Are not concerns of this sort determined by forces possessing vast inertia, and thus powerful sweep, that can at most be slightly dislodged from their predestined course but definitely not arrested or completely turned around?

In the first lecture I indicated that public power in the United States would not have spread to the present level if we had rationalized our choice and had limited our considerations to ideological angles. Under properly implemented and executed operating policies, private enterprise can perform as well as, and can perhaps outperform, any publicly owned electric power operation. An examination of public operation in such countries as France, England, Italy, Mexico, and Russia gives no basis for questioning that view. That in spite of all this public power in the United States has spread in a half-century some sixfold must be accepted as clear testimonial to the most probable reason. Private enterprise did not perform according to its potentials and failed, or was judged to have failed, in performing its functions of supplying acceptable service to its public constituents.

Have there been such failures? The answer is not many, but perhaps enough failures to have left a deep imprint, and collectively perhaps a cumulative effect also. For example, the private utilities in their early history made a great botch in their policies toward the lighting of their cities and thus brought into being scores of municipal operations. Rural electrification was another area of failure. The REA came into being essentially because the industry failed to comprehend its manifest duty to bring electricity to the producers of the food of the people of the United States. The failure to interconnect thoroughly, necessity for which was evident more than 30 years ago, resulted from inability to visualize the catastrophic effect of complete breakdown of an isolated system. The failure to insure reliability through coordinated planning, designing, and building of main system facilities was caused by exaggerated notions of independence and sovereignty.

In connection with atomic power, in recent years the industry made two grave errors. The first was equating the placing of an order for a reactor with a reactor placed in service and ready to carry load. The second was dropping coal without properly evaluating the transition period between atomic power and coal.

Currently the industry is in the midst of committing two more costly blunders: it fails to see the imperative of protecting its enriched nuclear fuel supply by taking over the three existing enrichment plants of the AEC and guaranteeing their subsequent expansion as becomes necessary. The second blunder is in not responding half as imaginatively as is possible, nor with half the necessary celerity, to the opportunity presented to it to take leadership in the solution of the pollution problem in every area connected with its operations: from particulate matter, SO_2, and the oxides of nitrogen; from thermal effects; from release of nuclear poisons; from noise; and from aesthetic pollution.

Perhaps these are not so many failures in a history that reaches back almost 90 years, particularly since the record of providing an adequate, highly economical, and highly reliable power supply has been, with a few minor exceptions, remarkably good and has been possibly the best provided in any modern society. But the industry cannot afford the risk of relying on having its failures excused on the ground that to err is human but to forgive divine.

What of the future? It is quite clear from what has already been said that the task ahead is Herculean, technologically, economically, organizationally. I believe that private organization, to an extent no less than prevails at the present time, would do better than public organization for the people of the United States; it is rational to believe that it will be entrusted with that responsibility, except for the evolutionary development of one force in our society about which I have so far remained silent.

Our society, operating essentially on the principle of private ownership, provides economic goods of two kinds—individual and social. Individual goods are divisible, and each person or household buys particular objects and individual services on the basis of his choice as a consumer. Social goods are not divisible into individual items or possessions. They are communal services—national defense, education, flood control, park services, elimination of pollution, slum elimination, abolition of poverty, and they can neither be sold nor bought by individuals or adjusted to individual tastes. The nature, the amount, and the cost of these goods must be set by a single decision applicable jointly to all persons. Social goods, therefore, are subject to communal or political decision rather than to individual demand.[15]

It is a paradox that, as our private enterprise system grows in size, complexity, and in production and productivity, and as energy becomes more and more interwoven with our daily living, there is a tendency to look on energy, and particularly on electric energy, more and more as a social operation and less as an economic one. This is where the danger lies to those concerned with the continuation of private ownership and management of our power supply facilities. Too many of the managers of our utility enterprise have been raised by and large in a background of law, or accounting, or economics. Technical imagination is not part of their training. Social responsibility and ethics are more or less foreign concepts—at least they would be treated as improper for integration into the management decision process. Even the concept of dedication to public service has missed many of them. There now exists a movement among

[15] For further discussion, see Report of the National Commission on Technology, Automation, and Economic Progress, "Technology and the American Economy," vol. 1 (Washington, D.C.: U.S. Government Printing Office, February 1966).

many of the utilities free from SEC supervision to seek greener pastures and enter diverse fields of activity, including airlines, computers, housing, prefabricated homes, plastics for toys, motels, and all kinds of similar operations possessing one characteristic in common: the outlook and approach of all are alien to a sound utility outlook and approach.[16]

What the utility industry needs is a much greater number of managers capable of being fascinated by the complexity and technological interest and business challenge presented by the power industry, not only for developing a highly profitable and consumer-satisfying operation but for participating actively in the many areas of social contact and in the efforts to reorient the industry with a view of facing up to the problems and challenges I have indicated. Unless they do, the industry it seems to me is in very grave danger; the apparent stabilization point of 23.3% that we have seen develop between public and private power will be upset, and power by private enterprise will decline regardless of the clear logic for its continuation.

This would be a sad development. It seems to me that the collective drive of all the data and discussion of the four lectures I am about to complete is toward the conclusion that there is no social form of organization which can cope with, and magically eliminate, all difficulties in the electric utility industry in its dynamic growth and development. Indeed, that is equally true of society generally.

The resolution of our difficulties requires, above

[16] "Greener Pastures?" *The Wall Street Journal,* 27 July 1970.

all, integrity and competence of a high order rather than organizational change. And yet it is also clear that power by private enterprise may continue to decline until it meets some more potent resistance point. Could it be at 50% (±2.5%) as in the Swedish power systems? It could be.

With a change in the tax laws to put *all* power production and distribution on a par taxwise, with the elimination of financing subsidies, and with regulation at all levels extended to the public power sector, a truly competitive relationship between public and private power could be brought about. Under such a condition the arrest of the extension of public power at a roughly fifty/fifty division line could serve a useful and constructive socioeconomic purpose: Private power operation could serve as a true yardstick for public power and vice versa. This would be an ironic but much less cynical use of the yardstick concept.

As I pointed out in my first lecture, the Swedish power supply is a highly successful integration of public and private power, and the regulations under which the two segments operate not only gain for the state important sources of revenue but keep alive ethical and real competition between the public and private sectors. That integration, therefore, might not be as much a catastrophe as complete public power would be. But the inevitability of such change will be forestalled if the private enterprise segment recognizes the threat and the challenge and responds with alacrity, imagination, vigor, and effectiveness.

INDEX